Contents

Executive Summary

Growing awareness of the built environment's impact on the natural environment, economy, health, and productivity has spurred rapid growth in the green building industry. Green buildings maximize operational efficiencies while minimizing negative environmental and health impacts.

LEED: DEFINING LEADERSHIP FOR NEW AND EXISTING BUILDINGS

In 2000, USGBC established the Leadership in Energy and Environmental Design (LEED) Green Building Rating System™ as a way to assess sustainable achievements for the built environment. LEED certification is available for both new and existing buildings as well as neighborhoods. For new construction projects, owners can design and construct healthy, high-performance buildings right from the start.

Greening existing buildings, on the other hand, may require system upgrades, retrofits, installations, or renovations, as well as the implementation of operations and maintenance (O&M) best practices and sustainable policies. Many owners want to green their existing buildings, but often perceive the needed improvements to be cost prohibitive.

LEVERAGING
SAVINGS TO PAY
FOR GREEN UPGRADES

The paid-from-savings approach is a financing strategy to green existing buildings. It leverages the savings generated from building system upgrades to pay for a comprehensive greening project within a defined pay-back period. Paid-from-savings projects can use a variety of financing methods including:

- Self-financing,
- tax-exempt lease-purchase agreements for qualifying entities,
- power purchase agreements for renewable energy projects,
- performance contracts for larger projects,
- equipment finance agreements, and
- commercial loans or bond financing for qualifying entities.

In many cases, successful projects employ a combination of these options, along with supplemental funding, such as revolving loan funds, utility rebates, and renewable energy grants, as well as funds from the organization's capital and operating budgets.

Using the paid-from-savings approach allows owners to implement needed repairs and upgrades, achieve reductions in energy and water use, and incorporate other green strategies and technologies in the most cost-effective manner.

This overview provides basic information to help building owners understand the paid-from-savings approach and decide if it is a viable option to green their existing building, including the steps to assess if the building has the potential to achieve LEED certification. Project profiles illustrate the variety of project types suited to this approach.

WHAT IS LEED?

LEED is an internationally recognized certification system that measures how well a building performs using several metrics, including:

- energy savings,
- water efficiency,
- CO_2 emissions reduction,
- improved indoor environmental quality, and
- stewardship of resources.

The rating systems provide a concise framework for identifying and implementing practical and measurable green building design, construction, operations, and maintenance solutions.

LEED points are awarded on a 100-point scale, and credits are weighted to reflect their potential environmental impacts. A project must satisfy specific prerequisites and earn a minimum number of points to be certified. Certification levels, based on the number of points, are: Certified, Silver, Gold, and Platinum.

LEED for Existing Buildings: Operations & Maintenance

The LEED for Existing Buildings: Operations & Maintenance rating system is a set of performance standards for the sustainable ongoing operation, maintenance, and retrofit of buildings that are not undergoing major renovations. It addresses high-performance building systems, O&M best practices, and sustainable policies.

LEED Offers a Full Suite of Green Building Rating Systems

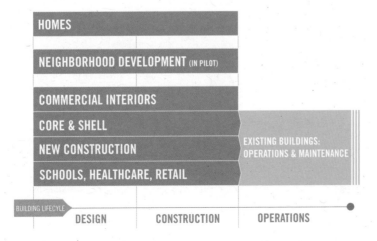

HOMES

NEIGHBORHOOD DEVELOPMENT (IN PILOT)

COMMERCIAL INTERIORS

CORE & SHELL

NEW CONSTRUCTION

SCHOOLS, HEALTHCARE, RETAIL

EXISTING BUILDINGS: OPERATIONS & MAINTENANCE

BUILDING LIFECYLE

DESIGN   CONSTRUCTION   OPERATIONS

The LEED for Existing Buildings: O&M rating system can be applied both to existing buildings seeking LEED certification for the first time and to projects previously certified under LEED for New Construction, Schools, or Core & Shell. It is the only LEED rating system under which buildings are eligible for recertification.

To some degree, all efforts to install high-performance building systems to lower energy and water use and to reduce greenhouse gas emissions will improve a building's environmental performance. The focus of this guide, however, is to assist owners in applying the paid-from-savings approach to seek LEED for Existing Buildings: O&M certification.

Value of LEED

LEED is a smart business decision. As the project profiles illustrate, installing high-performance building systems can yield significant utility cost savings. A LEED-certified building also showcases an owner's commitment to the environment and demonstrates an understanding of how a green work environment improves occupant productivity and health.

Paid-from-Savings
Approach to
Project Funding

The paid-from-savings approach leverages cost savings generated from building system upgrades to pay for a comprehensive greening project within a defined pay-back period. The cost-saving measures can vary regarding installation costs and pay-back periods. They include such items as:

- replacing the boiler,
- replacing the chiller,
- upgrading lighting systems,
- installing a building automation system (BAS), and
- replacing water fixtures.

Owners can achieve their desired return on investment (ROI) and lessen the overall project pay-back period by "bundling" the longer pay-back measures with the quicker pay-back measures to create a project with a shorter overall pay-back period and a higher ROI.

The Best Candidates for a Paid-from-Savings Approach

In general, the best candidates for the paid-from-savings approach are buildings with inefficient or outdated building systems in which upgrades will generate significant cost savings. To achieve LEED for Existing Buildings: O&M certification, these systems must also meet energy-efficiency and performance-period requirements designated in the rating system. LEED for Existing Buildings: O&M certified buildings also implement O&M best practices and sustainable policies. Successful LEED projects require the commitment of the owner to ensure these practices and policies are adopted and maintained.

Performance Contracting as a Project-Delivery Method

The two common methods for delivering a paid-from-savings project are 1) the traditional renovation and retrofit installation process and 2) performance contracting (PC), which is a project-delivery method that includes a financing strategy. Performance contracting is a well-established means of procuring needed building repairs and upgrades. It focuses on building system upgrades that yield utility and other operating cost savings. Among the factors that may influence the selection of the project-delivery method are the size and scope of the project, staff expertise, and cost. Financing can be included in the performance contract or provided by independent third party financial institutions.

ESCO Guarantees the Savings

Under a performance contract, an energy services company (ESCO) acts as the project developer and assumes the technical and performance risk associated with the project, including guaranteeing the cost savings generated from the system upgrades for a specified period of time. If the savings guarantee is not met, the ESCO pays the owner the difference. The guarantee is unique to performance contracting and not typical of other paid-from-savings approaches.

To determine the savings that can be guaranteed, the ESCO will conduct an investment-grade energy audit, which provides the basis for calculating the guarantee and creating the project development plan. The audit also serves as the foundation for developing the measurement and verification (M&V) plan, which outlines the specific methods and calculations to ensure the expected savings are realized.

Comparing Performance Contracting to Green Performance Contracting

Traditional performance contracting will improve a building's environmental performance by installing high-performance building systems that reduce energy and water consumption. Green performance contracting (Green PC) is based on the same project-delivery method as traditional performance contracting, but enhances the process by utilizing the LEED for Existing Buildings: O&M rating system as the criteria for a comprehensive green project. The range of measures in a Green PC project is broader than the utility-system upgrades found in traditional performance contracting. While Green PC is designed to facilitate LEED certification, the ESCO cannot guarantee it, as many of the credits, especially those related to O&M best practices and sustainable policies, fall under the purview of the owner, not the ESCO.

Key Steps of a
Paid-from-Savings Project Seeking LEED Certification

The following is a general outline of the steps related to a paid-from-savings project seeking LEED for Existing Buildings: O&M certification. The process is fluid and the timing for completing the steps will vary based on project specifics, such as the building's condition, staff capacity, financing needs and availability, state laws and regulations, and project economics. The steps are a roadmap, providing an overview of the process and the tasks involved.

STEP 1
Understand LEED Requirements

- Review Minimum Program Requirements (MPRs) for LEED for Existing Buildings: O&M and ensure the building is a viable candidate for LEED for Existing Buildings: O&M certification.

STEP 2
Project Preparation

- Form a project team with organizational stakeholders.
- Ensure team members are familiar with the LEED for Existing Buildings: O&M rating system and the paid-from-savings approach.
- Determine the building's current energy performance rating using the EPA's ENERGY STAR® Portfolio Manager tool.
- Evaluate the potential for cost savings using EPA's Cash Flow Opportunity (CFO) Calculator.

STEP 3
LEED Certification Assessment

- Conduct the LEED certification assessment to determine if the building will, upon completion of upgrades, meet the nine LEED for Existing Buildings: O&M prerequisites.

STEP 4
Project Economics Assessment and Financing

- Through an energy and water auditing process and the LEED certification assessment, project measures are identified.
- Owner and the project team can modify the list of project measures to ensure all desired LEED credits are included.
- Determine project financing needs (dollar amount, terms, potential borrowing limitations, etc.) and research options.
- Determine how the building's utility systems will be maintained to ensure savings are continually generated.
- Match the M&V process needed to the level of financial risk.
- Assess any potential cash-flow problems.
- Ensure the identified project measures — including those required for LEED certification — are bundled* to create the desired ROI and simple pay-back period.

STEP 5
Project Implementation

The project-delivery method will define the implementation process.

- For projects using a traditional renovation or retrofit process:
  - Establish M&V procedures to ensure cost savings are realized.
  - Implement building system improvements, O&M best practices and sustainability policies.
  - Manage the LEED documentation process.
- For Green PC projects:
  - Select an ESCO.
  - Negotiate an agreement with the ESCO.
  - Conduct an investment-grade energy audit.
  - Establish an M&V Plan.
  - Finalize the Green PC Agreement.
  - Implement building improvements.
  - Manage the LEED documentation process. ESCO may assist with LEED credit implementation and documentation.

STEP 6
LEED Certification

- Ensure performance period requirements are met. Nearly all LEED for Existing Buildings: O&M prerequisites and credits have a performance-period requirement that begins when all requirements are fully implemented and functioning.
- Manage documentation process using LEED Online.
- Submit LEED certification application to Green Building Certification Institute (GBCI) for review at the end of the performance period.

bundled :: In paid-from-savings projects, building system improvements generate utility cost savings. These savings are leveraged to help fund the project. Paid-from-savings projects seeking LEED certification can "bundle" or aggregate the utility cost-saving measures with non cost-saving measures to optimize green opportunities and project economics. When longer pay-back measures are combined with the quicker measures, the project will have a shorter overall pay-back period and higher ROI.

Additional
Implementation
Steps for Green PC Projects

The following is a general outline of the additional steps related to a paid-from-savings project using Green PC as the project-delivery-method. The steps are a roadmap, providing an overview of the process and the tasks involved.

STEP A Green PC Preparation

- Ensure team members know the fundamentals of performance contracting and the laws and regulations that govern it in the state.
- Conduct a PC mini-audit to assess whether the project meets the criteria for the performance contracting project-delivery method.

STEP B ESCO Partner Selection

- Develop the RFP/Q; indicate the goal of achieving LEED for Existing Buildings: O&M certification.
- Determine the ESCO selection process; require the ESCO to be knowledgeable of the LEED for Existing Buildings: O&M rating system.
- Add the ESCO to the project team when selected.

STEP C Investment-Grade Energy Audit

- Include LEED certification assessment tasks in the audit's scope-of-work.
- Data collected sets the basis for the savings guarantee, project development plan, and M&V plan.

STEP D Project Development Plan

- ESCO identifies proposed project measures as savings opportunities and LEED certification opportunities.
- Final plan will bundle all project measures— including those needed for LEED certification— to ensure the desired ROI and simple pay-back period are realized.

STEP E Measurement & Verification Plan

- Describe the pre- and post-project conditions, based on the energy audit, and how these conditions will generate savings.
- Describe how actual savings will be verified.
- Owner and the ESCO agree on how the adjusted baseline will be calculated.

STEP F Green PC Agreement

- Modify traditional performance contract language to include details on efforts to seek LEED certification.
- Outline the responsibilities of the ESCO and those of the owner.

Summary

The LEED for Existing Buildings: O&M rating system contains clearly defined performance targets, yet, as the project profiles illustrate, the path to implementation can be flexible. Owners can develop a financing package using the cost savings from system upgrades, along with a host of other options, including rebates, grants, revolving loan funds, tax credits/incentives, and equipment lease agreements.

The paid-from-savings project illustrated in the chart below has been simplified to help demonstrate how project measures are bundled to reach a desired ROI and simple pay-back period. Most projects will have a more complex financial analysis that encompasses the costs and potential savings from all credits pursued for LEED certification, including any recurring operating costs associated with efforts to ensure recertification.

Green Performance Measures for Existing Buildings (LEED prerequisite/credit)	Capital Budget Costs	Operating Budget		
		Onetime Costs	Annual Costs	Annual Savings
High Performance Building Systems				
Plant Native Plants & Groundcover (SSc5)		$8,250		$400
Install Water Efficient Fixtures (WEp1, WEc2)	$22,000			$2,190
Install Energy Efficiency Improvements (EAp2, EAc1)	$505,473			$70,375
Test & Balance Outside Air Intakes (IEQp1)		$21,250		
O&M Best Practices				
Occupant Commuting Survey (SSc4)		$0	$0	
Develop Landscape Plan & Training (SSc3)		$2,500		
Conduct ASHRAE Level II Audit (EAc2.1)		$17,000		
Implement Low/No-cost Improvements and On-going Cx (EAc2.2, EAc2.3)		$18,500	$1,500	$8,500
Conduct Waste Stream Audit (MRc6)		$6,250		
Conduct IAQ Audit (IEQc1.1)		$8,750		
LEED Assessment & Documentation Services		$25,500		
Sustainable Policies				
Develop Sustainable Purchasing Policy/Program (MRp1)		$3,750	$2,250	$0
Develop Recycling Policy/Program (MRp2, MRc7)		$6,750	$1,000	$2,680
Establish ETS Control Policy (IEQp2)		$0	$0	
TOTALS	$527,473	$118,500	$4,750	$84,145
	Total Cost: $645,973		Net Savings: $79,395	
ROI	12.3%			
PAYBACK	8.1 Years			

Green project measures can be tailored to project specifics, such as the condition of the building, state laws and regulations, the skill level of in-house staff, and the project provider's expertise. Using the paid-from-savings approach to green an existing building, owners save on utility costs, improve the building's asset value, help the environment, and create a work environment that improves occupant productivity and health.

Success Stories

Many institutions and corporations have used a paid-from-saving approach to green existing facilities and achieve LEED certification.

The following profiles provide details on successful projects and illustrate the variety of project types suited to the approach, including a convention center, a 100-year-old state capitol building, and a million-square-foot corporate headquarters. A university's innovative revolving loan fund is also included.

- Adobe Systems Headquarters
- California EPA (Cal/EPA)
- Colorado State Capitol Complex
- Dallas Convention Center
- Harvard University's Green Capital Loan Fund (GCLF)
- National Geographic Society Headquarters

Adobe Systems Headquarters
San Jose, CA

Background

Since its founding in 1982, Adobe Systems, Inc. has strived to be an environmentally friendly company. Located in downtown San Jose, CA, Adobe occupies one million square feet of commercial office space. In 2001, the California energy crisis spurred Adobe to review the energy and water efficiency of its headquarters. As a result, the company implemented new building systems, established operations and maintenance best practices, and adopted green policies.

Adobe was awarded LEED for Existing Buildings Platinum Certification for its headquarters complex in 2006, and for two regional headquarters buildings located in San Francisco. Adobe was the first commercial enterprise to achieve a total of four LEED platinum certifications, solidifying its reputation as a leader in promoting environmental stewardship and creating healthy work environments.

Photo courtesy of: William Porter

Green Performance Measures and Cost Savings

Adobe's investment of $2.1 million in energy and environmental retrofits are saving $1.5 million in energy and water costs annually. On a per square foot basis, electricity use has been reduced by 39% and solid waste diverted by 98%.

Green performance measures included upgrading chillers and retrofitting the main supply fans with variable frequency drives. Interior lighting systems were retrofitted, timers on garage exhaust fans and outdoor lighting systems were installed, and sensors were added to monitor carbon monoxide levels. Adobe increased its use of outdoor air and enhanced the overall maintenance of its air systems, resulting in better indoor air quality. The company also implemented a green cleaning program.

Water usage was reduced 38% by installing flow restrictors on all faucets, low-flow shower heads, and waterless urinals. Site irrigation was reduced 76% by planting drought-tolerant landscaping and installing a drip-irrigation system with eT-controllers, which adjust landscape watering automatically according to real-time conditions communicated from local weather stations using wireless technology.

Paid-from-Savings Approach

Adobe used a paid-from-savings approach to finance the implementation of green performance measures. It primarily self-financed the project, however, preceding LEED certification in 2006, Adobe received a total of $389,000 in rebates from various sources, including $350,000 from its local utility for participation in energy efficiency programs. Adobe also received rebates directly from the California Public Utilities Commission and $5,000 from the city of San Jose to fund water conservation measures. These rebates reduced the total cost to Adobe from $2.1 million to $1.7 million. Adobe's paid-from-savings approach yielded a return on investment of 91% with a simple payback period of 1.1 years.

KEY PROJECT FACTS

Project size: 1 million SF
Cost: $1.7 million
Reductions: electricity – 39%
water – 38%
diverted solid waste – 98%
Annual savings: $1.5 million
Simple payback period: 1.1 years

LEED for Existing Buildings Platinum.

Photo courtesy of: Walter Drane

Background

In 2004, the California Environmental Protection Agency's (Cal/EPA) Joe Serna Jr. Building was the first building to receive Platinum LEED for Existing Buildings Certification. Located in downtown Sacramento, Cal/EPA occupies the largest high-rise building in the city with 25 stories and 950,000 square feet. Cal/EPA's pioneering efforts to be the first to achieve LEED Platinum certification demonstrates the agency's commitment to its mission to protect and restore California's natural resources.

Green Performance Measures and Cost Savings

Green performance measures included the installation of highly efficient HVAC and lighting systems, photovoltaic rooftop panels and low-mercury lighting tubes with perimeter light sensors to automatically dim lights when natural sunlight is sufficient. As a result of these efforts and others, the Cal/EPA building consumes 60% less energy per square foot than other high-rise buildings in its district and boasts an ENERGY STAR® rating of 99 out of a possible 100 points. After an initial investment of $3.5 million in green performance measures, the annual savings are $1.6 million.

Operating costs have been lowered dramatically through reduced water usage and waste disposal. Low-flow toilets, waterless urinals, and water-efficient fixtures have decreased exterior water use by 50% and interior water use by 20%. Use of a vermicomposting system diverts more than 10 tons of waste from landfills a year, saving $10,000 annually. This unique system uses 20,000 to 30,000 red wiggler worms to devour Cal/EPA's café food prep waste. In addition, requiring re-usable cloth garbage bags instead of garbage can liners, saves tens of thousands of dollars a year.

Performance Contracting

In addition to the green performance measures described above, Cal/EPA entered a performance contract with an energy services company to implement a groundwater project. A new filtration system significantly reduces the minerals in the water increasing the efficiency of the cooling system and decreasing municipal water use and electricity needed to pump water to the building. If monthly savings do not meet the guarantee, the energy services company pays Cal/EPA the difference. The project was implemented at zero additional cost to taxpayers.

KEY PROJECT FACTS

Project size: 950,000 SF; 25 floors

Cost: $3.5 million

Reductions: exterior water use – 50%
interior water use – 20%

Annual savings: $1.6 million

Simple payback period: 2.2 years

LEED for Existing Buildings Platinum.

Colorado State Capitol
Denver, CO

Background

Built in 1895, the Colorado State Capitol may not appear to be a likely candidate for LEED certification, but its success proves historic buildings can be models of sustainability.

In 2008 the Capitol building achieved LEED for Existing Buildings: Operations & Maintenance certification during a comprehensive greening of the 20 buildings within the Capitol complex.

Photo courtesy of: Colorado Governor's Energy Office

Green Performance Measures and Cost Savings

Green performance measures across the complex included modifying the chilled water system, replacing the cooling tower, and installing energy-efficient chillers, irrigation controls, and low-flow toilets. Traditional cathode ray tube (CRT) computer screens were replaced by energy saving light-emitting diode (LED) screens, and all computer systems are turned off at night. As a result of these efforts and others, the complex has reduced energy consumption by 34%. The sustainable policies implemented include a recycling program, a green cleaning policy, use of eco-friendly landscaping products and plans, and an employee education program that encourages staff to conserve energy and resources.

Performance Contracting

The costs to improve the complex were $24 million, including $900,000 to renovate the Capitol building. To finance the Capitol complex project, the state entered a 19-year performance contract with an energy services company. The contract guarantees $1.1 million in annual energy savings from the improvements to the Capitol complex.

In 2005, Colorado began using the criteria outlined in the LEED for Existing Buildings rating system for its state measurement and verification (M&V) process. The M&V process ensures equipment is maintained and the energy and water systems are generating the anticipated cost savings.

KEY PROJECT FACTS

Project size: 1.6 million SF; 20 buildings
Cost: $24 million
Reductions: energy – 34%
Annual savings: $1.1 million
Simple payback period: 21.8 years

Colorado State Capitol
LEED for Existing Buildings:
Operations & Maintenance Certified.

Dallas Convention Center
Dallas, TX

Background

Occupying 2.2 million square feet in the downtown central business district, the Dallas Convention Center (DCC) demonstrates how even the largest of buildings can successfully pursue LEED for Existing Buildings certification using a paid-from-savings approach.

Constructed in 1957, the DCC is composed of 105 meeting rooms, two ballrooms, a 1,740-person theater, a 75 berth truck loading dock and almost a million square feet of exhibit space. Although registered in 2006, the DCC was further inspired by a 2008 city-wide commitment to draw 40% of its energy use from renewable sources. These efforts proved a catalyst for DCC to implement additional green performance measures and seek LEED certification.

Photo courtesy of: Dallas Convention Center

Green Performance Measures and Cost Savings

Green performance measures included building retrofits focused on optimizing performance and minimizing waste. To reduce energy usage, 30-year-old chillers were replaced with new high-efficiency models with variable frequency drives. The DCC also replaced more than 30,000 high-mercury content bulbs with T5 and T8 compact florescent bulbs, which have helped to position the DCC to save 20 million kilowatt hours of electricity per year. Fifty-four solar-thermal collection panels to heat water were installed and power capacitors on five electrical services were added, increasing the power factor efficiency to 95%. By replacing plumbing fixtures with low-flush models and installing a more efficient cooling tower, the DCC reduced water consumption by seven million gallons or 18% annually.

In its pursuit of LEED certification, the DCC has implemented a purchasing policy that requires caterers to use biodegradable cups and plates. The DCC has also committed to recycling 50% of its waste over five years, and has developed an education program to help clients and area hotels develop sustainability policies.

Performance Contracting

In October 2008, the City of Dallas entered a ten-year performance contract with an energy services company. The ESCO replaced most of the facility's lighting, faucets and flush valves, and the cooling tower. It also installed two high-efficiency chillers and new motors and pumps. Under the performance contract, the ESCO has guaranteed utility savings of $2 million annually and will continue to offer measurement and verification (M&V) services and operational support.

Paid-from-Savings Financing Strategy

In addition to the performance contract, the DCC financed the project through a $16 million loan and funds from its operating budget. The DCC also partnered with an electrical provider who agreed to provide a rebate for electrical efficiencies once the improvements had been monitored and verified over a 12-month period. The DCC's paid-from-savings project had a simple payback period of eight years.

KEY PROJECT FACTS

Project size: 2.2 million SF
Cost: $16 million
Reductions: energy – 35%
water – 18%
Annual savings: $2 million
Simple payback period: 8 years

Currently seeking LEED for Existing Buildings certification.

Background

Renowned for its academic excellence, Harvard University has also established itself as an institution of environmental excellence. The Green Campus Loan Fund (GCLF), a $12-million revolving loan fund to promote sustainability improvements to Harvard's campus, was created to bridge gaps in the capital and operating budgets. With maximum payback criteria of five or ten years, depending on the loan type, projects funded through the GCLF have averaged a 27% return on investment (ROI), producing over $4 million in savings.

Photo courtesy of: Harvard Office for Sustainability

Green Campus Loan Fund

The GCLF offers a variety of methods to fund proposed greening projects. Applicants may choose a full-cost loan, which covers the entire cost of a project limited to $500,000 per green measure and with a payback period of no more than five years, or an incremental loan, which offers a maximum of $500,000 per green measure for the cost delta between standard efficiency equipment and premium efficiency equipment with an internal rate of return of 9% or higher. GCLF also funds feasibility studies for renewable energy loans and enhanced metering loans.

Rockefeller Hall

Rockefeller Hall is a successful paid-from-savings project funded, in part, by the GCLF. Originally opened in 1971 as a student residence and community center, Rockefeller Hall was a gathering place for Harvard Divinity School students and a refectory prior to its June 2007 renovation. The renovation included green performance measures, such as new lighting controls, CO_2 sensors to manage ventilation, a sun-bouncing white roof to reduce cooling costs, and an energy recovery wheel to save energy by regulating seasonal heat and moisture exchange between indoor and outdoor air.

A gear-driven elevator saves up to 40% in energy costs compared to a typical hydraulic elevator. The renovations are saving $22,000 in energy costs and 75 metric tons of CO_2 annually. The expansive nature of the Rockefeller Hall renovations qualify the project to seek certification under the LEED for New Construction rating system.

KEY PROJECT FACTS

Investment: $12 million revolving loan fund

Return on investment: project average of 27% annually

Loans distributed in first seven years: $11.5 million

Rockefeller Hall is currently seeking LEED for New Construction certification.

National Geographic Society Headquarters
Washington, DC

Background

Established in 1888, the National Geographic Society (NGS) is a world-renowned, non-profit educational and scientific institution. Its headquarters complex is located in Washington, DC, and is comprised of three, inter-connected class A commercial buildings, totaling 840,000 square feet. The oldest was constructed in 1902; the newest in 1984.

The society's goal for seeking LEED certification was to operate facilities that reflected its mission while remaining cost effective. The society's success was acknowledged by receiving the first LEED for Existing Buildings Silver certification in November 2003.

Photo courtesy of: The National Geographic Society

Green Performance Measures and Cost Savings

Green performance measures included replacing chillers and boilers and adding air-handling systems, window film, a white roof, energy-efficient lighting, and an energy management control system. NGS also implemented water conservation measures including efficient flush valves, and indoor air quality projects such as upgrading the building's management system controls for CO_2 monitoring and improved temperature controls. As a result of the renovations and upgrades, energy consumption is 2.5 million kilowatt hours less than it was in 1996, water usage is down 18%, and waste costs have been reduced by 70%.

NGS also established green practices related to landscaping, site maintenance, construction waste management, snow removal, and pest management. It increased bike storage capacity and implemented policies encouraging telecommuting and the use of hybrid vehicles.

Demonstrating its continued support of sustainability, NGS created a "Go-Green Steering Committee" in 2006, which monitors the company's commitment to green performance and suggests additional sustainability policies and practices. At the committee's recommendation, all computers are now automatically shut off at 10:00 p.m. to conserve electricity. NGS is considering installing light emitting diode (LED) lighting.

Performance Contracting

Project renovations were financed and managed in part through a performance contract with an energy services company. The $1.8 million performance contract included HVAC system improvements tied to a total guaranteed energy savings of 8%.

Paid-from-Savings Financing Strategy

Total project costs were financed through a Washington, D.C. revenue bond including the $1.8 million performance contract.

KEY PROJECT FACTS

Project size: 3 buildings, 840,000 SF
Cost: $6.5 million
Reductions: energy – 8-11%
　　　　　　waste – 70%
　　　　　　water – 18%
Annual savings: $406,000
Simple payback period: 16 years

LEED for Existing Buildings:
Operations & Maintenance Silver.

Introduction

The U.S. Green Building Council's (USGBC) *Paid-from-Savings Guide to Green Existing Buildings* provides owners, architects, engineers, facilities managers, and energy service companies (ESCOs) with detailed information on how to use the paid-from-savings approach to green an existing building with the goal of achieving LEED for Existing Buildings: Operations & Maintenance certification.

The paid-from-savings approach is a financing strategy to green existing buildings. It leverages the savings generated from building system upgrades to pay for a comprehensive greening project within a defined pay-back period. Paid-from-savings projects can use a variety of financing methods including:

- Self-financing,
- tax-exempt lease-purchase agreements for qualifying entities,
- power purchase agreements for renewable energy projects,
- performance contracts for larger projects,
- equipment finance agreements, and
- commercial loans or bond financing for qualifying entities.

In many cases, successful projects employ a combination of these options, along with supplemental funding, such as revolving loan funds, utility rebates, and renewable energy grants, as well as funds from the organization's capital and operating budgets.

A project's cost-saving measures can vary regarding installation costs and pay-back periods. They include such items as: replacing the boiler, upgrading lighting systems, installing building automation system (BAS) controls, and replacing water fixtures. Owners can achieve their desired return on investment (ROI) and lessen the overall project pay-back period by bundling* the longer pay-back measures with the quicker measures to create a project with a shorter overall pay-back period and a higher ROI.

bundled :: In paid-from-savings projects, building system improvements generate utility cost savings. These savings are leveraged to help fund the project. Paid-from-savings projects seeking LEED certification can "bundle" or aggregate the utility cost-saving measures with non cost-saving measures to optimize green opportunities and project economics. When longer pay-back measures are combined with the quicker measures, the project will have a shorter overall pay-back period and higher ROI.

Just as owners have multiple options for financing the project, there are several project delivery methods that can be applied to paid-from-savings projects. These include the traditional renovation and retrofit process, in which a project manager oversees the work of in-house staff and/or, green performance contracting. Factors influencing the method chosen will include the size and scope of the project, financing needs and availability, staff expertise, and project economics.

Performance Contracting

Performance contracting — also known as energy performance contracting (EPC) or guaranteed energy savings performance contracting (ESPC) — is a well-established means of procuring and financing building repairs and upgrades. It is both a paid-from-savings financing strategy and a project-delivery method that focuses on energy and water system upgrades that yield significant utility cost savings.

Under performance contracting, building owners contract with an ESCO to act as the project developer for a wide range of tasks and to assume the technical and performance risk associated with the project, including guaranteeing the cost savings generated from the system upgrades. As with all paid-from-savings projects, these costs savings are used to pay for the upgrades within a specific pay-back period.

Green Performance Contracting (Green PC)

Green PC is based on the same project-delivery methods as traditional performance contracting, but enhances the process by utilizing the LEED for Existing Buildings: O&M rating system as the criteria for a comprehensive green project. To some degree, all efforts to install high-performance building systems with the goal of lowering energy and water use and reducing greenhouse gas emissions will improve the building's environmental performance. However, the range of project measures in a Green PC project is broader than the building system upgrades found in a traditional performance contract, adding additional measures that can increase savings and further reduce the building's environmental footprint.

Successful Green PC projects require two essential processes — performance contracting and LEED certification — to be integrated and managed simultaneously. Green performance contracting:

- Provides more in-depth evaluation of utility cost-saving measures to ensure the minimum levels of energy- and water-efficiency performance required for LEED certification are met, and
- Includes green project measures that may not generate cost savings, but contribute toward the project's goal of LEED certification.

Owners and project teams must determine if the project meets the criteria for green performance contracting and if the LEED credits can be accomplished through the Green PC process. An

ESCO cannot guarantee LEED certification because many of the credits, especially those related to operations and maintenance best practices and sustainable policies, fall under the purview of the owner. However, an ESCO can serve as the LEED Coordinator and can manage the LEED documentation process if the owner so chooses.

The Green PC process is fluid, although the timing for completing the steps will vary based on project specifics, such as the building's condition, staff capacity, financing needs and availability, state laws and regulations, and project economics. The second section of the guide details the additional preparation and implementation steps needed for Green PC and provides technical information for integrating LEED requirements into traditional performance contracting. The steps that follow are a roadmap, providing an overview of the process and the tasks involved.

The Paid-from-Savings Approach to Green Existing Buildings

Regardless of the project-delivery method chosen — the traditional renovation process or green performance contracting — the first steps in developing a paid-from-savings project seeking LEED certification are the same and include assessments to determine if the project has the potential to:

- generate significant cost savings from the building system upgrades to make the project economically feasible, and
- achieve the nine prerequisites required for LEED for Existing Buildings: O&M certification, including the critical energy-efficiency performance requirements.

If the project is deemed viable for the paid-from-savings approach, the project-delivery method will dictate the next step, the implementation phase. The final step for projects seeking LEED certification includes managing the LEED documentation process and submitting the final application to the Green Building Certification Institute (GBCI) for review.

What is LEED?

In 2000, USGBC established the LEED (Leadership in Energy and Environmental Design®) rating system as a way to assess sustainable achievements for the built environment. It is an internationally recognized certification system that measures how well a building performs across all the metrics that matter most: energy savings, water efficiency, CO_2 emissions reduction, improved indoor environmental quality, and stewardship of resources and sensitivity to their impacts. LEED provides building owners and operators a concise framework for identifying and implementing practical and measurable green building design, construction, operations, and maintenance solutions.

The suite of LEED rating systems are designed to address the complete lifecycle of buildings. LEED rating systems address new and existing commercial, institutional, and residential buildings and include:

- New Construction
- Existing Buildings: Operations & Maintenance
- Commercial Interiors
- Core & Shell
- Schools
- Retail
- Healthcare
- Homes
- Neighborhood Development

LEED® Rating System

- Sustainable Sites
- Water Efficiency
- Energy & Atmosphere
- Materials & Resources
- Indoor Environmental Quality
- Innovation in Operations
- Regional Priority

All are based on accepted energy and environmental principles and strike a balance between established practices and emerging concepts.

Each rating system is organized into credit categories: Sustainable Sites, Water Efficiency, Energy and Atmosphere, Materials and Resources, Indoor Environmental Quality, and Innovation in Operations. Regional bonus points are another feature of LEED and acknowledge the importance of local conditions in determining best environmental design and operations practices.

LEED points are awarded on a 100-point scale, and credits are weighted to reflect their environmental impacts. A project must satisfy all prerequisites and earn a minimum number of points to be certified. Certification levels, based on the number of points, include: Certified, Silver, Gold, and Platinum.

Once the LEED credits are implemented and the energy-efficiency and performance requirements met, the final step for LEED certification is submitting the project certification documentation. When all templates are submitted and all required supporting documents uploaded to the Web-based LEED Online system, the application is ready to submit for review. The Green Building Certification Institute (GBCI) reviews applications for LEED certification and provides feedback. If the application is complete and all requirements are met, GBCI awards a LEED certification for the building.

LEED for Existing Buildings: Operations & Maintenance

The LEED for Existing Buildings: O&M rating system helps owners and facilities managers measure operations, improvements, and maintenance on a consistent scale to maximize operational efficiency while minimizing environmental impacts. It addresses system upgrades, whole-building cleaning and maintenance practices, recycling programs, and exterior maintenance efforts. The rating system can be applied both to existing buildings seeking LEED certification for the first time and to projects previously certified under LEED for New Construction,

Schools, or Core & Shell. It is the only LEED rating system under which buildings are eligible for recertification.

LEED for Existing Buildings O&M Credit Categories

Sustainable Sites (SS) Credits address building systems and maintenance activities related to a building's exterior and site. Programs to reduce the use of automobiles, heat island effect, and light pollution are included.

Water Efficiency (WE) Credits address plumbing fixtures and fittings, water usage, landscape irrigation systems, and cooling tower water management.

Energy and Atmosphere (EA) Credits address the building's energy performance, building commissioning, refrigerant management, energy-use monitoring, and emissions-reduction reporting.

Materials and Resources (MR) Credits address sustainable purchasing and solid waste management.

Indoor Environmental Quality (IEQ) Credits address outdoor air ventilation, indoor air quality, occupant comfort, and green cleaning.

Innovation in Operations (IO) Credits provide the opportunity to earn points for additional environmental benefits not addressed in the rating system. Points can be attained by achieving exemplary performance of an existing credit or by implementing an operation, practice, or upgrade not outlined in the rating system.

Project Measures Addressed Within LEED for Existing Buildings: O&M

High-Performance Building Systems: implement building and site improvements and technologies in order to use less energy, less water, and fewer natural resources. System upgrades and retrofits also improve indoor environmental quality and address operational inefficiencies. Examples: Efficient lighting systems, indoor plumbing fixtures, and ecologically appropriate features.

O&M Best Practices: adopt operations and maintenance best practices to ensure project measures are effectively implemented and maintained. Examples: Systems monitoring, green cleaning, and preventative maintenance procedures.

Sustainable Policies: establish green policies, programs, and plans to demonstrate an organization-wide commitment to sustainability. Examples: Recycling programs and the use of eco-friendly products.

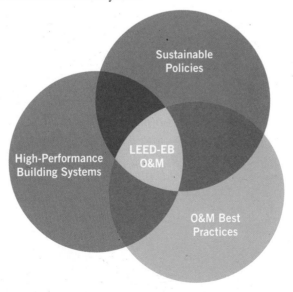

LEED Prerequisites and Credits and Their Potential to Generate Savings

Tables outlining all LEED for Existing Buildings: O&M prerequisites and credits are located in **Appendix B**. Each prerequisite and credit is identified as either having a high potential to generate savings or a having the potential to impact savings. Prerequisites and credits that will yield no savings are also noted. From the tables, owners can garner a sense of the types of credits that can help generate the cost savings required for a paid-from-savings project. In most cases, the opportunities for savings are found in the prerequisites and credits related to building system improvements and upgrades. The final project measures of a paid-from-savings project seeking LEED certification will not be chosen solely for their potential to generate savings, but by other factors as well, including the level of certification pursued, the overall project economics, the skill levels of the facilities and procurement staffs, state laws and regulations, the project provider's expertise, and the owner's requirements.

Sample Paid-from-Savings Project

Figure I-1 is an example of a paid-from-savings project that achieved LEED for Existing Buildings: O&M Gold Certification. All prerequisites and credits in the rating system are listed and the possible points for each credit are noted. (Points are not awarded for prerequisites.) Project scenarios will vary, but this example is representative of the types of credits pursued in a paid-from-savings project. In this case, the project earned 61 points out of 110 for Gold Certification. The example also illustrates that the implementation of some project measures will rest with the owner, such as adopting O&M best practices and sustainable policies. The completion of other measures, particularly those associated with building system improvements and upgrades, will be outsourced to third parties.

Figure I-1. Sample Paid-from-Savings Project

SUSTAINABLE SITES	Points Pursued / Possible Points	Outsourced	In-house
SSc1 – LEED Certified Design and Construction	0/4	[Not pursued]	[Not pursued]
SSc2 – Building Exterior and Hardscape Management Plan	1/1		Create a plan
SSc3 – Integrated Pest Management, Erosion Control, and Landscape Management Plan	1/1		Create a plan
SSc4 – Alternative Commuting Transportation	10/15		Conduct survey
SSc5 – Site Development – Protect or Restore Open Habitat	1/1		Landscape plantings
SSc6 – Stormwater Quantity Control	1/1		Assess site
SSc7.1 – Heat Island Reduction – Nonroof	0/1	[Not pursued]	[Not pursued]
SSc7.2 – Heat Island Reduction – Roof	1/1	Roof replacement	
SSc8 – Light Pollution Reduction	1/1	Interior, exterior light fixtures (full cutoff)	
SUSTAINABLE SITES – TOTAL POINTS	16/26		

Figure I-1. Sample Paid-from-Savings Project (continued)

WATER EFFICIENCY	Points Pursued / Possible Points	Outsourced	In-house
WEp1 – Minimum Indoor Plumbing Fixture and Fitting Efficiency		Plumbing fixture replacements	
WEc1 – Water Performance Measurement	1/2	Install water meters/ submeters	
WEc2 – Additional Indoor Plumbing Fixture and Fitting Efficiency	3/5	Plumbing fixture replacements	
WEc3 – Water Efficient Landscaping	3/5	Irrigation system	
WEc4 – Cooling Tower Water Management	0/2	[Not pursued]	[Not pursued]
WATER EFFICIENCY – TOTAL POINTS	**7/14**		

ENERGY AND ATMOSPHERE	Points Pursued / Possible Points	Outsourced	In-house
EAp1 – Energy Efficiency Best Management Practices – Planning, Documentation, and Opportunity Assessment		Energy audit, BOP, SO, SN analysis	
EAp2 – Minimum Energy Efficiency Performance		Efficiency measures	
EAp3 – Fundamental Refrigerant Management		Refrigerant conversion	
EAc1 – Optimize Energy Efficiency Performance	10/18	Efficiency measures	
EAc2.1 – Existing Building Commissioning – Investigation and Analysis	2/2	ASHRAE Level II Energy Audit	
EAc2.2 – Existing Building Commissioning – Implementation	2/2	Install measures	
EAc2.3 – Existing Building Commissioning – Ongoing Commissioning	0/2	[Not pursued]	[Not pursued]
EAc3.1 – Performance Measurement – Building Automation System	1/1	System installation	
EAc3.2 – Performance Measurement – System-Level Metering	1/2	System installation	
EAc4 – On-site and Off-site Renewable Energy	0/6	[Not pursued]	[Not pursued]
EAc5 – Enhanced Refrigerant Management	0/1	[Not pursued]	[Not pursued]
EAc6 – Emissions Reduction Reporting	1/1	Provide report	
ENERGY AND ATMOSPHERE – TOTAL POINTS	**17/35**		

MATERIALS AND RESOURCES	Points Pursued / Possible Points	Outsourced	In-house
MRp1 – Sustainable Purchasing Policy			Create policy
MRp2 – Solid Waste Management Policy			Create policy
MRc1 – Sustainable Purchasing – Ongoing Consumables	1/1		Tracking system
MRc2 – Sustainable Purchasing – Durable Goods	1/2		Tracking system
MRc3 – Sustainable Purchasing – Facility Alterations and Additions	1/1	Project materials specification	
MRc4 – Sustainable Purchasing – Reduced Mercury in Lamps	1/1	Order, install	
MRc5 – Sustainable Purchasing – Food	0/1	[Not pursued]	[Not pursued]
MRc6 – Solid Waste Management – Waste Stream Audit	1/1		Conduct audit
MRc7 – Solid Waste Management – Ongoing Consumables	0/1	[Not pursued]	Bins, tracking system
MRc8 – Solid Waste Management – Durable Goods	1/1		Tracking system
MRc9 – Solid Waste Management – Facility Alterations and Additions	0/1	[Not pursued]	[Not pursued]
MATERIALS AND RESOURCES – TOTAL POINTS	**6/10**		

Figure I-1. Sample Paid-from-Savings Project (continued)

INDOOR ENVIRONMENTAL QUALITY	Points Pursued / Possible Points	Outsourced	In-house
IEQp1 – Minimum Indoor Air Quality Performance		Design, install, measure, test	
IEQp2 – Environmental Tobacco Smoke (ETS) Control			Establish policy
IEQp3 – Green Cleaning Policy			Create policy
IEQc1.1 – IAQ Best Management Practices – IAQ Management Program	1/1		Conduct audit, adopt protocols
IEQc1.2 – IAQ Best Management Practices – Outdoor Air Delivery Monitoring	1/1	Design, install	
IEQc1.3 – IAQ Best Management Practices – Increased Ventilation	0/1	[Not pursued]	[Not pursued]
IEQc1.4 – IAQ Best Management Practices – Reduce Particulates in Air Distribution	1/1	Design, install	
IEQc1.5 – IAQ Best Management Practices – IAQ Management for Facility Alterations and Additions	1/1	Construction activity specification	
IEQc2.1 – Occupant Comfort – Occupant Survey	1/1		Conduct survey
IEQc2.2 – Controllability of Systems – Lighting	1/1	Design, install	
IEQc2.3 – Occupant Comfort – Thermal Comfort Monitoring	1/1	Design, install	
IEQc2.4 – Daylight and Views	0/1	[Not pursued]	[Not pursued]
IEQc3.1 – Green Cleaning – High-Performance Cleaning Program	1/1		Create program
IEQc3.2 – Green Cleaning – Custodial Effectiveness Assessment	0/1	[Not pursued]	[Not pursued]
IEQc3.3 – Green Cleaning – Purchase of Sustainable Cleaning Products and Materials	1/1		Adopt practice
IEQc3.4 – Green Cleaning – Sustainable Cleaning Equipment	1/1		Adopt practice
IEQc3.5 – Green Cleaning – Indoor Chemical and Pollutant Source Control	1/1		Adopt practice
IEQc3.6 – Green Cleaning – Indoor Integrated Pest Management	1/1		Implement plan
INDOOR ENVIRONMENTAL QUALITY – TOTAL POINTS	**12/15**		

INNOVATION IN OPERATIONS	Points Pursued / Possible Points	Outsourced	In-house
IOc1 – Innovation in Operations	1/4		Use of alternative fuels
IOc2 – LEED® Accredited Professional	1/1	Consultant's staff	
IOc3 – Documenting Sustainable Building Cost Impacts	0/1	[Not pursued]	[Not pursued]
INNOVATION IN OPERATIONS – TOTAL POINTS	**2/6**		

REGIONAL PRIORITY CREDITS	Points Pursued / Possible Points	Outsourced	In-house
RPc1 – Regional Priority	1/4	Met threshold for WEc2	
REGIONAL PRIORITY CREDITS – TOTAL POINTS	**1/4**		

	Points Pursued / Possible Points	Certification Level	
TOTAL POINTS	**61/110**	**Gold**	

The Paid-from-Savings Approach to Green Existing Buildings

STEP 1
Understand LEED Requirements

STEP 2
Project Preparation

STEP 3
LEED Certification Assessment

STEP 4
Project Economics Assessment and Financing

STEP 5
Project Implementation

STEP 6
LEED Certification

Owners need to determine if LEED for Existing Buildings: Operations & Maintenance is the appropriate rating system for the proposed project. To decide if a building is eligible for certification, owners need to consider the following:

1. Does the building meet the minimum program requirements?
2. Can the building meet the required performance periods?
3. Will the alterations and additions follow the rating system's guidelines?
4. Will the building meet the nine prerequisites and achieve the minimum points required for certification upon completion of the building system upgrades?

The criteria to help answer these four questions are outlined below. Projects that do not qualify may be eligible for certification under the LEED for New Construction, Schools, or the Core & Shell rating systems.

1. Does the building meet the minimum program requirements?

The LEED 2009 Minimum Program Requirements (MPRs) define the minimum characteristics that a project must possess in order to be eligible for certification under LEED 2009. These requirements define the categories of buildings that the LEED rating systems were designed to evaluate, and taken together, serve three goals: to give clear guidance to customers, to protect the integrity of the LEED program, and to reduce the challenges that occur during the LEED certification process. A full list of MPRs can be found in the "LEED 2009 for Existing Buildings Operations and Maintenance Rating System".

2. Can the building meet the required performance periods?

Some credits in the LEED for Existing Buildings: O&M rating system require that performance data and other documentation be submitted for the performance period. The performance period is the specific, defined time interval for which sustainable operations performance is being measured. For projects seeking LEED for Existing Buildings: O&M certification for the first time, the performance period is the most recent period of operation preceding certification application and must be a minimum of three months for all prerequisites and credits except Energy & Atmosphere Prerequisite 2 and Credit 1, which have longer minimum durations. At the project team's option, the performance period for any prerequisite or credit may be extended to a maximum of 24 months preceding certification application. (For a building seeking recertification, the performance periods differ. Reference the *Green Building Operations and Maintenance Reference Guide, 2009 Edition* for more details.)

3. Do the alterations and additions follow the rating system's guidelines?

Facility Alterations and Additions

Although LEED for Existing Buildings: O&M certification focuses on sustainable ongoing building operations, it also embraces sustainable alterations and new additions to existing buildings. It should be noted that under the LEED for Existing Buildings: O&M rating system, alterations and additions have a specific meaning. The definition focuses on changes that affect usable space in the building. Alterations that affect no more than 50% of the total building floor area or cause relocation of no more than 50% of regular building occupants are eligible. Additions that increase the total floor area by no more than 50% are also eligible. Buildings with alterations or additions exceeding these limits should pursue certification under the LEED for New Construction or LEED for Schools rating system. Mechanical, electrical or plumbing system upgrades that involve no disruption to usable space are excluded.

4. Will the building meet the nine prerequisites and achieve the minimum points required for certification upon completion of the building system upgrades?

Buildings seeking LEED for Existing Buildings: O&M certification must meet the nine LEED prerequisites and achieve the minimum points required for certification. The prerequisites are detailed in Step 3 along with information on conducting a LEED certification assessment, which will help owners determine whether the building already complies with the prerequisites or will be able to upon completion of the greening improvements. The assessment is also used to decide which LEED credits to pursue.

The "LEED 2009 for Existing Buildings: Operations and Maintenance Rating System" is a valuable resource for owners and project team members to better understand the rating system and to assess a project's eligibility. It is available for free download on USGBC's Web site, www.usgbc.org.

Other LEED educational and training opportunities are outlined in Step 2 of this guide.

STEP 1
Understand LEED
Requirements

STEP 2
Project
Preparation

STEP 3
LEED Certification
Assessment

STEP 4
Project Economics
Assessment and
Financing

STEP 5
Project
Implementation

STEP 6
LEED
Certification

The project preparation phase includes assembling the project team, education and training, and assessment activities. Many owners and project team members may not be familiar with the paid-from-savings approach and the LEED for Existing Buildings: O&M certification process. A clear understanding of both processes is needed to ensure the project's success.

Workloads and time constraints of in-house staff may require the owner to outsource some or all of the project preparation and assessment efforts. Third-party sustainability consultants or ESCOs can be contracted to assist with the up-front analysis.

A. Assemble the Project Team

The project team is composed of organizational stakeholders — maintenance and procurement staff, facilities managers, architects and engineers, business managers, and others — who will be responsible for due diligence and project management. The team will be involved in all aspects of the project, including the development of project measures and the LEED certification process. For Green PC projects, ESCOs should be added to the team when selected. This integrative approach to the project team is essential to the success of the green building project as it ensures owners, occupants, and operators understand their role throughout the lifecycle of the building.

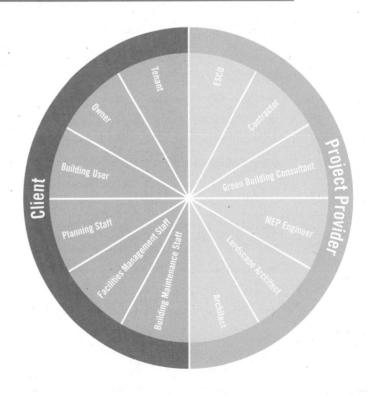

B. Accomplish Pre-Project Education Activities

- Visit USGBC's Web site for information on the LEED for Existing Buildings: O&M rating system.

- Attend workshops or Webinars on the LEED for Existing Buildings: O&M rating system.

- Research financing options.

- Gather sample contract documents and templates.

- Talk to peers in other organizations about their experiences with paid-from-savings projects.

For Projects Considering Green PC

- Visit the Energy Services Coalition's (ESC) Web site, www.energyservicescoalition.org, for information on performance contracting.
- Attend workshops or Webinars on performance contracting.

- Attend informal presentations by ESCOs.

- Consider third-party consulting services to help organize project documents, plan and implement the ESCO selection process, and develop the procedures to review contracts.

Additional Web site addresses and resources can be found in **Appendix C.**

C. Accomplish Pre-Project Assessment Activities

Several important assessment and evaluation activities should be performed before the project starts:

• Benchmark with ENERGY STAR® Portfolio Manager

Projects should determine the building's current energy performance rating using the EPA's ENERGY STAR® Portfolio Manager, the required benchmarking platform for validating a building's energy performance in the LEED for Existing Buildings: O&M rating system. Portfolio Manager is an interactive, online tool that allows the owner to track and assess energy and water consumption, performance, and cost information for individual buildings and building portfolios. Based on monthly utility data entered into the online tool, Portfolio Manager rates the current level of energy performance. If the existing building is one of EPA's eligible space types, it receives an energy performance rating on a scale of 1 to 100, known as the ENERGY STAR rating. If the building is not eligible for an ENERGY STAR rating, Portfolio Manager is still used to obtain the building's annual weather-normalized Source Energy Use Intensity. This value is then entered into the USGBC "Case 2" calculator to determine the energy performance.

ENERGY STAR is a joint program of the U.S. Environmental Protection Agency and the U.S. Department of Energy; www.energystar.gov

The rating may be the first indication of how feasible it will be to meet the minimum energy efficiency performance required by the LEED for Existing Buildings: O&M rating system. A building must achieve an ENERGY STAR rating of at least 69 (or at least 21% better than the national average source energy intensity for the non-ratable space types) to be eligible for LEED for Existing Buildings: O&M certification. In general, buildings with ENERGY STAR ratings below 50 are good candidates for the paid-from-savings approach because the building system upgrades have the potential to generate significant energy and water cost savings for the project.

- ## Utilize the Cash Flow Opportunity (CFO) Calculator

  EPA's ENERGY STAR CFO Calculator helps decision-makers answer three critical questions about energy efficiency investments: (1) how much new energy efficiency equipment can be purchased from the anticipated savings, (2) should the equipment purchase be financed now, or is it better to wait and use cash from a future budget, and (3) is money being lost by waiting for a lower interest rate? The CFO Calculator is an EXCEL® spreadsheet-based tool that allows owners to enter building performance data from Portfolio Manager and determine how much of the investment needed could be paid from current and future energy savings from a building systems upgrade. The tool lets owners analyze how various project financing scenarios will affect the project's financial performance.

  Using CFO Calculator in the early stages of project development is helpful because project team members learn the potential utility cost savings, which can be used to gain support for the project from hesitant stakeholders. The CFO calculator is free and can be found on the ENERGY STAR Web site.

- ## Prepare for the LEED Certification Assessment

  The purpose of a LEED certification assessment is to determine early in the project development phase the feasibility of attaining LEED for Existing Buildings: O&M certification. It focuses on the building's potential to meet the nine LEED prerequisites, including the critical energy efficiency requirements. Preparing for the LEED certification assessment includes assembling required project data and information and determining who will conduct the assessment. For project's using the traditional renovation and retrofit process, the assessment can be completed in-house or by third-party consultants. For Green PC projects, the owner can complete the assessment prior to developing the RFP/Q or can include it as part of the ESCO's scope-of-work. The LEED certification assessment is detailed in Step 3.

- ## Prepare for the Project Economics Assessment

  An economic assessment for a paid-from-savings project will:

    - Estimate the range of utility cost savings that each project measure can generate.
    - Establish if the measures, in aggregate, can achieve the desired return on investment (ROI) and simple pay-back period.
    - Determine the financial mechanism that will support the project's cash flow requirements.

  Preparing for the assessment includes deciding how the energy and water cost savings will be determined. Options include using empirical project data or Building Information Modeling (BIM). For Green PC projects, the ESCO will conduct an investment-grade energy audit using building energy modeling to ascertain the cost-saving potential. The Project Economics Assessment is detailed in Step 4.

Information on how to calculate a project's ROI and simple pay-back period are outlined below. Owners often focus on the ROI of an individual project measure and not the overall project ROI. This limited approach may result in the elimination of important project measures because they do not, on their own, produce the desired ROI.

ROI

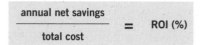

To calculate the ROI, divide the *annual net savings* by the *project's total cost*. The ROI will vary depending on both the size of the project and the amount of savings produced.

Simple Pay-Back Period

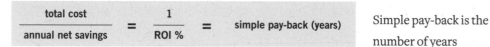

Simple pay-back is the number of years it will take to recover installation costs based on annual cost savings. To calculate the simple pay-back period, divide the *project's total cost* by the *annual net savings*. The simple pay-back period is the mathematical inverse of the ROI. To determine a target for simple pay-back, match the ROI to the project life. For example, if the project life is 12 years, then the matching ROI — the mathematical inverse of simple pay-back — would be 8.3%. Bundling quicker pay-back measures, such as lighting upgrades, with longer pay-back measures, such as chillers and large HVAC components, helps lower the overall pay-back period.

• Clarify Project Expectations

Owners who have experience with the LEED for New Construction rating system should expect to be more involved in a project seeking LEED for Existing Buildings: O&M certification. In addition to the building system improvements to meet energy-efficiency requirements, other owner-led measures must also be accomplished, such as adopting and implementing O&M best practices and sustainable policies. Several project scenarios emerge when considering options for a paid-from-savings project seeking LEED certification.

- **Turnkey Project Scenario** – Certification is an outcome of the paid-from-savings project. Many owners will have this scenario in mind when contemplating LEED certification in conjunction with a paid-from-savings project. Clearly defined roles and responsibilities will help owners and project providers establish reasonable expectations for the timing and accomplishment of all deliverables. Early education and understanding of the LEED rating system will be a more significant project preparation step.

- **Owner Leads the Way Scenario** – In this scenario, the owner has already accomplished a number of LEED credit requirements through the adoption of sustainable policies and the implementation of O&M best practices. To achieve certification, all that remains is to improve the building's physical systems to ensure the performance requirements are met. This is an ideal project scenario since the owner is already knowledgeable about the LEED rating system and, therefore, is in a better position to specify the project provider's scope-of-work. Certification can be readily facilitated using the paid-from-savings approach to fund buildings system improvements.

- **Project Leads the Way Scenario** – This scenario will be typical of owners who are new to the process of greening existing buildings. The owner is committed to pursuing LEED certification and wants to ensure the building system improvements will meet the standards required by the rating system. Remaining owner-led measures, such as policy adoption and O&M best practices implementation, are not expected to be completed at the time the building system improvements of the paid-from-savings project are implemented. However, the owner is committed to a time table that includes the completion of those measures at some point in the near future.

STEP 1
Understand LEED
Requirements

STEP 2
Project
Preparation

STEP 3
LEED Certification
Assessment

STEP 4
Project Economics
Assessment and
Financing

STEP 5
Project
Implementation

STEP 6
LEED
Certification

The LEED certification assessment is the process to evaluate the project's potential to achieve the nine prerequisites required for LEED for Existing Buildings: O&M certification. The prerequisites, listed below, are categorized as high-performance building systems, O&M best practices, and sustainable policies. Assessment checklists can be found in **Appendix D**.

The project team can perform the assessment using in-house staff or third-party consultants. For projects using Green PC, if there are doubts as to whether the building system improvements will meet the minimum LEED energy efficiency requirements, the owner can ask the ESCO to make the determination. See Green PC Audit Agreement in **Appendix I**.

LEED for Existing Buildings: O&M Prerequisites

High-performance Building Systems

The assessment of the building system prerequisites requires measuring a building's energy efficiency performance, water usage, and outside air ventilation. Energy efficiency performance is assessed using Portfolio Manager. Water usage can be estimated by establishing the building's age, inventorying existing plumbing fixtures, and completing required calculations. Evaluating outside air ventilation, rates requires modifying or maintaining each air handing unit with an outside air connection.

1. Minimum Energy Efficiency Performance (EA Prerequisite 2) focuses on the building's energy performance as determined by Portfolio Manager. Significant improvements in energy efficiency can be attained by retrofitting inefficient mechanical and lighting systems with newer and more efficient equipment. The owner must still determine whether the improvement, however significant, will meet the minimum energy performance rating required by the prerequisite, an ENERGY STAR rating of 69.

2. Minimum Indoor Plumbing Fixture and Fitting Efficiency (WE Prerequisite 1) covers minimum efficiency performance requirements for water plumbing fixtures and fittings based on the Uniform Plumbing Code of 2006.

3. **Minimum Indoor Air Quality Performance (IEQ Prerequisite 1)** concerns minimum outside air introduction based on the American Society of Heating, Refrigerating and Air-Conditioning Engineers (ASHRAE) Standard 62.1.

O&M Best Practices

The O&M best practices prerequisites require documentation of system operations, preventive maintenance activities, and the management of sustainable practices. The assessment process consists of reviewing documentation and performing site assessments/audits of O&M practices to ensure requirements are met.

4. **Energy Efficiency Best Management Practices — Planning, Documentation and Opportunity Assessment (EA Prerequisite 1)** involves gathering or establishing fundamental documents that serve as the foundation for commissioning activities related to energy and atmosphere credits and continued improvement in energy efficiency.

5. **Fundamental Refrigerant Management (EA Prerequisite 3)** involves the use of certain types of refrigerants in refrigeration systems.

6. **Environmental Tobacco Smoke (ETS) Control (IEQ Prerequisite 2)** deals with the isolation of tobacco smoke from interior spaces occupied by non-smokers.

Sustainable Policies

The assessment process for the sustainable policies prerequisites can be as simple as deciding to adopt such policies. Each policy must adhere to the LEED for Existing Buildings: Operations & Maintenance policy model. See LEED for Existing Buildings: O&M Policy Model in **Appendix E.**

7. **Sustainable Purchasing Policy (MR Prerequisite 1)** concerns the purchasing of eco-friendly consumable and durable goods.

8. **Solid Waste Management Policy (MR Prerequisite 2)** deals with the building's recycling program.

9. **Green Cleaning Policy (IEQ Prerequisite 3)** covers custodial practices, janitorial equipment, and use of green cleaning supplies.

Assessing Potential Improvement Measures

The LEED certification assessment addresses both the building's existing condition and its potential for improvement. The assessment identifies performance gaps and determines whether improvement measures will resolve any deficiencies. For most of the prerequisites, once a performance gap is identified, the resolution is clear. Assessing energy performance requirements, however, are more complex.

One particularly important prerequisite is Energy & Atmosphere Prerequisite 2—Minimum Energy Efficiency Performance. Once the current ENERGY STAR rating for the building is obtained, other steps are required to determine whether energy performance improvement measures will be enough to meet the minimum ENERGY STAR rating of 69. (Ideally, building owners will strive for the ENERGY STAR label, which requires a rating of at least 75.) Assessing the potential to achieve the required ENERGY STAR rating will also reveal the building system improvements needed.

Project teams pursuing LEED certification use EPA's Portfolio Manager to determine the building's current energy performance and generate an Energy Star rating using 12 months of actual utility data. To predict what level of energy use will be needed to achieve a desired rating, project administrators can use of the following approaches. The steps can be implemented by the owner or outsourced to a third-party.

Target Finder

EPA's Target Finder helps owners establish an energy performance target for building projects. It determines the energy use intensity (EUI), defined as kBtu/SF/year, that is needed for a building to attain a specified ENERGY STAR rating. The project team enters the required space attribute values, zip code, utility-cost information, and specifies a target rating — at least 69 for LEED certification. Target Finder then calculates the EUI for the desired rating. Target Finder and instructions for use can be found on the ENERGY STAR Web site.

To determine if the proposed project measures will reach the needed energy use intensity, the project team can use either energy modeling software or Building Information Modeling (BIM).

Energy Modeling

For Green PC projects, the ESCO will conduct an investment-grade energy audit to help determine the project measures needed to achieve the energy use intensity, as determined by Target Finder. The ESCO will use a building energy modeling software tool, such as a DOE 2-based modeling program, to determine the energy efficiency impact of proposed measures. The task can be written into the audit's scope-of-work and completed by the ESCO. If this option is selected, the energy performance assessment will not normally be completed until the ESCO is selected and the investment-grade energy audit completed. ESCOs also use energy modeling to determine the utility cost savings that the proposed measures can generate. For paid-from-savings projects using the traditional renovation and retrofit process as its project-delivery method, using energy modeling is often not necessary, since this level of detail on the project's energy savings value is not typically required.

Building Information Modeling (BIM)

BIM can be used to determine whether proposed measures will improve the building's energy efficiency enough to achieve the target ENERGY STAR rating. The BIM approach does not require as much time and effort as energy modeling. Therefore, using BIM in the pre-project planning stage is a reasonable alternative. The model can be created by the owner's in-house engineering staff or a third-party architectural/engineering firm. It could also be produced by a potential project provider during the project's pre-proposal stage. BIM captures key information about the facility, including location, building type, square footage, room volumes, occupancy capacity, equipment locations, weather data, and building envelope.

If the project team uses BIM, a second step is required to assess the building's potential energy performance. The BIM needs to be uploaded to an online service that interfaces with ENERGY STAR Target Finder, such as Green Building Studio®, which is an online fee-based model analysis service that uses an open-source platform. Green Building Studio enables BIM models from various proprietary software applications to assess the energy performance of proposed designs. A Target Finder interface within the Green Building Studio program identifies the potential ENERGY STAR rating that could be expected from the BIM being analyzed. The information is then used to determine the project's potential ENERGY STAR rating. See the BIM Overview in **Appendix K.**

STEP 1
Understand LEED
Requirements

STEP 2
Project
Preparation

STEP 3
LEED Certification
Assessment

STEP 4
Project Economics
Assessment and
Financing

STEP 5
Project
Implementation

STEP 6
LEED
Certification

Cost savings from improved energy efficiency is the single greatest opportunity for funding a paid-from-savings project. Energy efficiency is a fundamental element of all LEED projects. By redirecting a portion of future utility savings to cover the costs of financing other measures, projects teams can undertake comprehensive green improvements and achieve LEED certification in a cost-effective manner.

Components of a Paid-from-Savings Project:

- Project Development: defines the project measures that will generate utility cost savings.
- Ongoing Services: addresses how the equipment will be maintained over time to ensure savings are continually generated.
- Financing: describes how the underwriting is to be treated.

Project Development

The project development phase identifies the project measures through an energy and water auditing process and the LEED certification assessment. Owners and project team members can also make modifications and add credits according to their desired level of LEED certification. (For Green PC projects, the investment-grade energy audit will be used to identify project measures.)

The paid-from-savings project illustrated in **Figure 4-1** is based on an actual energy performance improvement project for a middle school that was self-financed by a school district in Florida. This project bundled measures to achieve an ROI of 12.3%. Project costs are differentiated between capital budget costs and operating costs. Net savings are calculated by subtracting annual costs from annual savings produced by the utility cost-saving measures.

Energy efficiency improvements increased the middle school's ENERGY STAR rating from 41 to over 70, surpassing the minimum requirements for the LEED energy performance prerequisite. In this example, the utility cost savings generated by the energy improvement measures helped pay for the other green performance measures resulting in a LEED certification.

The measures in the example represent the three types of project measures needed for LEED certification: high-performance building systems, O&M best practices, and sustainable policies. The costs for some of the prescribed measures can be estimated by obtaining quotes from service providers and systems manufacturers. Other estimates for improvements require a more rigorous assessment process, particularly the energy cost-saving measures. Assessment exercises like that shown in **Figure 4-1** can be used early in the project development phase to estimate the project's economic feasibility. As the project develops and more cost information becomes available, the actual numbers can replace the estimates to provide a more accurate financial picture.

For Green PC projects, the owner is responsible for implementing and maintaining O&M best practices and sustainable policies. However, ESCOs can be tasked to help secure these credits, as long as the risk of non-performance is equitably placed on the party that has the most control over its accomplishment. The ESCO's role can include guidance, assistance, training, providing policy or planning templates, or completing LEED-Online documentation requirements.

Figure 4-1. Sample Paid-from-Savings Project

Green Performance Measures for Existing Buildings	Capital Budget Costs	Annual Operating Budget		
		One Time Costs	Recurring Costs	Recurring Savings
HIGH PERFORMANCE BUILDING SYSTEMS				
Modification of Landscape Features to Include Native Plants (SSc5)		$8,250		$400
The measure is an example of a one-time cost that generated savings in reduced water usage. The restoration of native habitat to the project site is an additional benefit.				
Install Faucet Aerators and New Dual Flush Toilet Valves (WEp1, WEc2)	$22,000			$2,190
The measure demonstrates the significant cost savings from water fixture upgrades. The savings not only have a direct-cost benefit to the owner, they reduce the impact on the local water and wastewater treatment facilities.				
Install Energy Efficiency Improvement Measures (EAp2, EAc1): a. Upgrade Old Controls with new Building Automation System b. Install Variable Frequency Drives in 17 Air Handling Units c. Lighting System Retrofit d. Occupancy-based Lighting and HVAC controls e. Chiller Plant Primary/Secondary Loop Modifications to Correct Low Delta T	$505,473			$70,375
Energy efficiency measures were identified through the ASHRAE Level II Audit (Item 5).				
Test & Balance Outside Air Intakes and Exhaust Systems and Report (IEQp1)		$21,250		
O&M BEST PRACTICES				
Occupant Commuting Survey (SSc4)		$0	$0	
The measure incurred no costs, but understanding occupant commuting choices can illuminate opportunities for alternative commuting incentive programs.				
Develop Erosion Control, and Landscape Management Plan and Training (SSc3)		$2,500		
The measure is an example of a one-time cost to implement a best practice; no savings or additional recurring costs resulted. Benefits include opportunities for staff to learn best practices, decreased water use, and improved stormwater quantity control.				
Conduct ASHRAE Level II Audit (EAc2.1)		$17,000		
The ASHRAE Level II Audit identified several improvement measures. See "Install Energy Efficiency Improvement Measures" above and "Implement Low or No-cost Operational Improvements and Ongoing Commissioning Program" below. These measures lay the foundation for the on-going commissioning program. Implementation of these measures is the focus of EAc2.2 and maintenance of the measures after installation so that savings persist over time is the focus of EAc2.3.				
Implement Low or No-cost Operational Improvements and On-going Commissioning Program (EAc2.2 and EAc2.3) a. Program BAS to temperature guidelines b. Set up trend logging capability in BAS system c. Make adjustments to outside air dampers d. Install various sensors for systems monitoring e. Calibrate sensors annually f. Set up and conduct training on the ongoing commissioning program		$18,500	$1,500	$8,500
After the improvements of EAc2.2 have been implemented, an on-going commissioning program is established in order to maintain the improvements that were installed. The cost savings derived from on-going commissioning program accrue from ensuring that the effectiveness of measures put into place persists over time. Costs involved in this measure relate to the initial training, installing additional capabilities and features on the BAS for trending and systems monitoring and reporting, annual calibration of sensors, and development of an on-going commissioning guide for use by onsite maintenance technicians.				
Conduct Waste Stream Audit and Complete Report (MRc6)		$6,250		
Conduct IAQ Audit and Complete Report (IEQc1.1)		$8,750		
LEED Assessment & Submission Services		$25,500		
Depending on the owner's in-house capabilities and availability of time to devote to the effort, the LEED assessment, submission of LEED documentation, and management of LEED Online submittal templates may be contracted to a third-party consultant or project provider. The cost of this service will vary widely depending on the scope of services. Some of this cost may be attributed to recertification fees that will occur at least every five years if the owner wants to maintain the LEED certification.				

Figure 4-1. Sample Paid-from-Savings Project (continued)

Green Performance Measures for Existing Buildings	Capital Budget Costs	Annual Operating Budget		
		One Time Costs	Recurring Costs	Recurring Savings
SUSTAINABLE POLICIES				
Develop Environmental Preferable Purchasing (EPP) Policy and Training (MRp1)		$3,750	$2,250	$0
The measure created a recurring cost to the operating budget; however, the cost was bundled with other project measures to maintain an acceptable ROI. Benefits of implementing MRp1 include fewer toxins in the work place and reduced impact on natural resources and the local environment.				
Establish Recycling Program, Record Keeping Process, and Purchase Bins (MRc7)		$6,750	$1,000	$2,680
Initial onetime costs included the purchase of bins. The measure created a recurring cost to the operating budget; however, the recurring cost is offset by a much greater annual savings due to reduced service charges by the refuse company. The savings also help to recoup the initial cost of recycling bins.				
Establish ETS Control Policy (IEQp2)		$0	$0	
PROJECT TOTALS	$527,473	$118,500	$4,750	$84,145
TOTAL PROJECT ROI		12.3%		
SIMPLE PAY-BACK		8.1 years		

Figure 4-1 illustrates the cost implications of each measure of the paid-from-savings project. This example highlights why it is important to bundle the project measures in order to present the project's overall economic value, not just the benefits of one or two measures.

Calculate Net Annual Savings

gross annual savings ($84,145) — annual costs ($4,750) = $79,395

Calculate Total Project Costs

capital budget costs ($527,473) + onetime operating costs ($118,500) = $645,973

Calculate a Project's ROI

$$\frac{\text{annual net savings (\$79,395)}}{\text{total costs (\$645,973)}} = 12.3\%$$

Calculate a Project's Simple Pay-Back Period

$$\frac{1}{\text{ROI (.123)}} = 8.1 \text{ years}$$

When calculating ROI, owners and project providers need to keep in mind that some project measures have an added recurring cost, which needs to be deducted from the gross annual savings. In the context of a whole project — where several measures produce savings, some do not, and others add costs — the overall project produces economic value as well as other values not measured by utility cost savings.

Ongoing Services

Durability of the savings stream is critical to the success of a paid-from-savings project because the project's performance period can be as long as 20 years, therefore, vigilance in monitoring system performance is required.

Figure 4-2. Examples of Savings Durability Issues

Project Measure	Key Considerations
Upgrade Old Controls with New Building Automation Systems (BAS)	Building automation systems are only as good as the programming. Presumably, the new system will have features the old system did not, leading to an overall increase in the building's efficiencies. However, settings change, key sensors fall out of calibration, and actuators fail. Without consistent and thorough system inspections, the savings from the measure could degrade over time.
Install Variable Frequency Drives (VFDs) in 17 Air Handling Units	Ongoing maintenance and effective operations of VFDs are critical to maintaining the savings, which can degrade from lack of servicing by building operators.
Lighting System Retrofit	Lighting systems are fairly durable over time. Disrepair of lighting systems performance will not necessarily cause utility cost increases. However, if the operation of the lights becomes inefficient, such as lights left on during non-occupied hours, savings could be lost.
Occupancy-based Lighting and HVAC controls	Control systems hold significant promise when first installed. However, some of the systems may be vulnerable to override capabilities. If not controlled, the overrides can significantly reduce savings.
Chiller Plant Primary/Secondary Loop Modifications to Correct Low Delta T	The measure serves to correct an original faulty installation. Once corrected, the savings are fairly durable. However, the system should be inspected periodically to ensure controls, settings, and valves are functioning properly.

Financing

Even though a paid-from-savings project can generate the savings needed to finance a project over time, the savings are not often realized in the short term, which can cause cash-flow problems. If monthly progress payments are due to project providers, it is important to determine early in the process if cash flow will be an issue. If so, a financing mechanism, such as a construction loan or revolving loan fund, can address the "gap" and provide needed funding during the project's execution phase.

Figure 4-3. Cash-Flow Gaps In this example, the installation of equipment started in the project's second month, but the utility cost savings were not realized until the tenth month. In paid-from-savings projects, a financing mechanism can address the cash flow "gap" and provide needed funding during the project's execution phase.

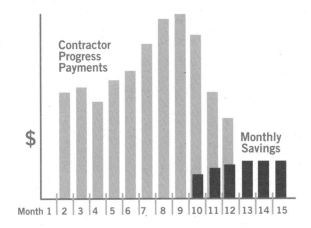

Financing Methods

Paid-from-savings projects can use a variety of financing methods including:

- tax-exempt lease-purchase agreements for qualifying entities,
- power purchase agreements for renewable energy projects,
- performance contracts for larger projects,
- equipment finance agreements, and
- commercial loans.

In many cases, successful projects employ a combination of these options, along with supplemental funding, such as revolving loan funds, utility rebates, renewable energy grants, federal and state tax incentives, as well as funds from the organization's capital and operating budgets. Details on these methods are in **Appendix F.**

Measurement & Verification (M&V) Process

Paid-from-savings projects with third-party financing need an added measure of assurance that system upgrades will perform. An M&V process will help mitigate the risk by outlining specific methods and calculations to ensure the expected savings are realized. The owner can decide the level of rigor needed for the M&V process based on the project's financial risk factors. For Green PC projects, the ESCO, to ensure the guaranteed savings are realized, will implement a formal M&V Plan based on data collected during the investment-grade energy audit and in consultation with the owner.

Owners using the traditional renovation and retrofit process to pursue LEED for Existing Buildings: O&M certification may not need an M&V Plan that is as detailed as that of a Green PC project, but nonetheless, may benefit from the components of such a plan. Key elements include outlining the procedures to measure the pre- and post-project conditions, detailing the methods on how to calculate the energy-use and adjusted baselines, creating a schedule for all M&V activities, and establishing O&M reporting responsibilities. Details on developing an M&V Plan are outlined in Step F and **Appendix G**.

Costs are associated with the M&V process; therefore, it is essential to develop the appropriate measurement method for each cost-saving improvement. Performance problems can arise if savings are "under-measured" and M&V costs can be unnecessarily high if savings are "over measured", so striking a balance is needed.

The International Performance Measurement and Verification Protocol (IPMVP) is the industry-standard process for M&V. The IPMVP provides a framework to determine energy and water savings that result from the implementation of an energy efficiency program. The Efficiency Valuation Organization (EVO) is a non-profit organization that develops the IPMVP products and services to aid in:

- Monitoring and verifying the energy and water savings that result from energy efficiency improvements.
- Managing the financial risk of energy savings performance contracts.
- Quantifying emissions reductions from energy efficiency projects.
- Promoting sustainable green, construction through cost-effective and accurate accounting of energy and water savings.

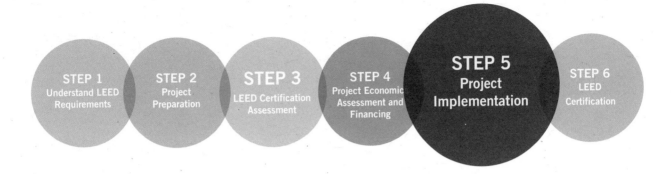

STEP 1
Understand LEED
Requirements

STEP 2
Project
Preparation

STEP 3
LEED Certification
Assessment

STEP 4
Project Economic
Assessment and
Financing

STEP 5
Project
Implementation

STEP 6
LEED
Certification

The project-delivery method will define the implementation process. The two common methods for a paid-from-savings project seeking LEED certification are the traditional renovation and retrofit process, in which a project manager oversees the work of in-house staff and/or contractors, and green performance contracting. Factors influencing the method chosen will include the size and scope of the project, financing, staff expertise, and cost.

For projects using a traditional renovation or retrofit process, additional tasks need to be added to the project's scope-of-work. The tasks include implementing O&M best practices and sustainable policies, establishing M&V procedures to ensure cost savings are realized, and managing the LEED documentation process.

For Green PC projects, all the elements of traditional performance contracting apply, including determining the ESCO selection process, negotiating an agreement, conducting an investment-grade energy audit, and establishing an M&V Plan. If detailed in the Green PC Agreement, the ESCO may also assist with implementing certain LEED credits and managing the LEED documentation process. The additional steps needed for Green PC projects are outlined in the section entitled Green Performance Contracting.

STEP 1
Understand LEED
Requirements

STEP 2
Project
Preparation

STEP 3
LEED Certification
Assessment

STEP 4
Project Economics
Assessment and
Financing

STEP 5
Project
Implementation

STEP 6
LEED
Certification

In paid-from-savings projects, building system improvements may be completed before all LEED prerequisites or credits have been implemented. The discrepancy in timelines is due in part to the fact that nearly all LEED prerequisites and credits require performance data to be collected over a performance-period requirement, which begins when all requirements for the prerequisite or credit are fully implemented and functioning. The end of all performance periods are when the LEED certification application is submitted to the Green Building Certification Institute (GBCI) for review. The minimum performance period is three months; the maximum performance period is two years.

LEED Online is the primary resource for managing the LEED documentation process. Through LEED Online, project teams manage project details, complete documentation requirements for LEED credits and prerequisites, upload supporting files, submit applications for review, receive reviewer feedback, and ultimately earn LEED certification. The Web-based LEED Online system provides a common space where members of a project team can work together to document compliance with the LEED rating system. With the exception of projects registered under LEED for Homes, all projects must be certified using LEED Online. GBCI reviews applications for LEED certification. If the application is complete and all requirements have been met, GBCI will award the building a LEED certification. For Green PC projects, the owner is responsible for the LEED documentation process and submitting the application to GBCI, unless these tasks are assigned to the ESCO in the Green PC Agreement.

The LEED for Existing Buildings: O&M rating system is the only LEED rating system under which buildings are eligible for recertification. Projects must apply for recertification within five years of the date of original certification, but may complete this process as frequently as every year. Maintaining policies, best practices, and record keeping after the original certification will allow owners to streamline the recertification process.

Green Performance Contracting (Green PC)

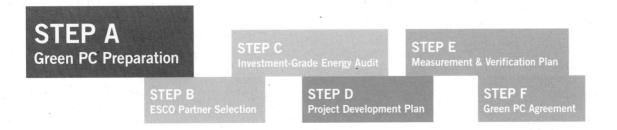

STEP A
Green PC Preparation

STEP B
ESCO Partner Selection

STEP C
Investment-Grade Energy Audit

STEP D
Project Development Plan

STEP E
Measurement & Verification Plan

STEP F
Green PC Agreement

The first section of the guide outlines the steps for implementing the paid-from-savings approach to achieve LEED for Existing Buildings: O&M certification for an existing building. The second half focuses on the additional steps needed when using green performance contracting as the project-delivery method. Green performance contracting utilizes the LEED for Existing Buildings: O&M rating system as the criteria for a comprehensive green project.

While many owners are familiar with the term performance contracting, they may not have experience using it to finance projects or as a project-delivery method. It is essential that owners have an understanding of the concepts and project tasks unique to performance contracting, including the statues and regulations that govern the process in the state in which the project is located, before moving forward. It is equally important to understand the requirements of the LEED for Existing Buildings: O&M rating system so that relevant requirements are included in the contracting and project documents. A list of education and information resources is available in **Appendix C**.

Many of the LEED requirements can be included in a performance contract so long as the risk of non-performance for the measure is equitably placed on the party—owner or ESCO—that has the most control over its accomplishment. For most LEED for Existing Buildings: O&M credits that address policies and practices, the owner will be responsible. The ESCO may be able to provide guidance, assistance, or training, but ultimately the owner needs to ensure policies and practices are implemented.

In preparation for a Green PC project, the owner should conduct a mini-audit to assess whether the project meets the criteria for the performance contracting project-delivery method. **Figure A-1** includes sample questions to consider when determining if a project is a good match for the performance contracting approach.

A-1. Sample Performance Contracting Mini Audit

	Yes	No	Assessment
1			Is the facility floor area greater than 150,000 square feet?
2			Are annual facility energy bills more than $200,000?
3			Does the building have aging equipment?
4			Are there recurring maintenance problems or high maintenance costs?
5			Are there comfort complaints?
6			Do maintenance budget constraints preclude major repairs?
7			Is there limited energy management expertise on staff?
8			Is the maintenance staff already overloaded with other tasks?
9			Have there been no major upgrades to the facility's lighting system?
10			Are major mechanical systems ready for replacement?

Courtesy of the Energy Services Coalition.

Green PC Documents

Owners and ESCOs will be working with several documents throughout the project. The documents for a Green PC project will be similar to those of a traditional PC project. However, the documents will have to be modified to align with and support the LEED certification effort. **Figure A-2** identifies the documents involved in a PC project. It also shows the relationships between the various documents. For example, the Project Development Plan is derived from information collected during the investment-grade energy audit in accordance with the audit agreement. All documents inform the final content of the Green PC Agreement.

Figure A-2. Green PC Modifications to Traditional PC

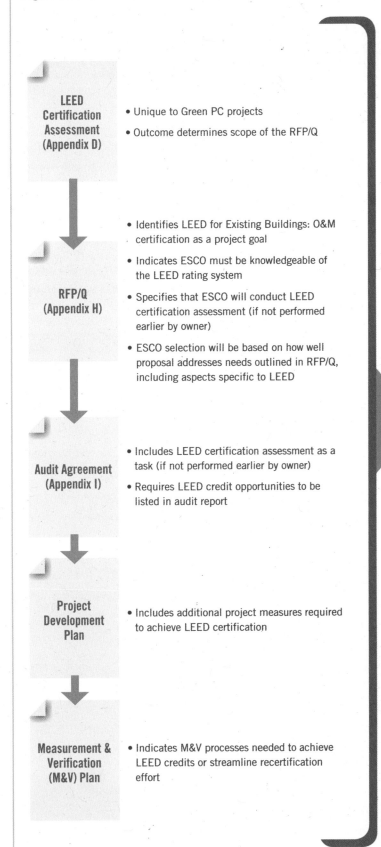

LEED Certification Assessment (Appendix D)

- Unique to Green PC projects
- Outcome determines scope of the RFP/Q

RFP/Q (Appendix H)

- Identifies LEED for Existing Buildings: O&M certification as a project goal
- Indicates ESCO must be knowledgeable of the LEED rating system
- Specifies that ESCO will conduct LEED certification assessment (if not performed earlier by owner)
- ESCO selection will be based on how well proposal addresses needs outlined in RFP/Q, including aspects specific to LEED

Audit Agreement (Appendix I)

- Includes LEED certification assessment as a task (if not performed earlier by owner)
- Requires LEED credit opportunities to be listed in audit report

Project Development Plan

- Includes additional project measures required to achieve LEED certification

Measurement & Verification (M&V) Plan

- Indicates M&V processes needed to achieve LEED credits or streamline recertification effort

Green PC Agreement (Appendix J)

- Modifies traditional Energy Services Agreement to include details on efforts to seek LEED certification
- If agreed that ESCO will assist with implementing LEED credits and/or services, specifies responsibilities
- References to earlier documents contained in specifications and schedules

STEP A	STEP C	STEP E
Green PC Preparation	Investment-Grade Energy Audit	Measurement & Verification Plan

STEP B	STEP D	STEP F
ESCO Partner Selection	Project Development Plan	Green PC Agreement

Some states have established procurement processes to support the owner's selection of an ESCO. There may also be statutes or regulations on what can be included in a performance contract.

The RFP/Q needs to include notification that modifications will be made to the standard performance contracting process in order to achieve LEED for Existing Buildings: O&M certification, and that the successful ESCO needs to be familiar with the LEED rating system. See Green PC RFP/Q Language in **Appendix H**. Further details on the RFP/Q process, as well as sample documents, can be found on the Energy Services Coalition Web site.

Selection Activities Include:

1. Conduct Due Diligence

Third-party consultants are available to conduct the necessary due diligence, which includes assistance with developing the selection process, evaluating proposals, and reviewing contracts. For owners with little or no experience with performance contracting, this approach is recommended. In such cases, owners should seek LEED Accredited Professionals (LEED APs) with LEED for Existing Buildings: O&M project experience.

2. Develop an RFP/Q

Owners will need to customize the RFP/Q to include the goal of achieving LEED certification. Credentials relating to LEED, such as the need for a LEED Accredited Professional to be on the ESCO project team or requests for examples of involvement on previous LEED for Existing Buildings: O&M projects, should also be detailed in the RFP/Q.

3. Provide Detailed Project Information to the ESCOs

ESCOs will need information about the facility before submitting an RFP/Q response, so project team members should be prepared to provide copies of utility bills, as-built drawings, BAS trend logs, and maintenance and service records. An invitation should be extended to interested ESCOs to visit the project site. The project teams should also inform the ESCOs of the sustainability goals or the specific LEED credits the owner wants to pursue. If the project team completed a LEED certification assessment prior to selecting the ESCO, the results should be made available. Some ESCOs may perform a LEED certification assessment as part of the proposal development process.

4. Evaluate Proposals

Owners should assess the qualifications of each ESCO based on specific skills and experience needed for the project. In addition to the written proposals, owners may want each ESCO to give a presentation and to provide client references. Experience with the LEED certification process should be a key evaluation criterion, and the team should ask for case studies or examples of any prior Green PC or LEED for Existing Buildings: O&M project experience.

Green PC projects require a guaranteed cost-saving stream over a defined period of time. To ensure the analysis of potential savings is accurate, the ESCO will perform an investment-grade energy audit.

The Investment-Grade Energy Audit Provides:

1. Basis for the Guarantee

Data collected during the investment-grade energy audit forms the basis for the energy savings guaranteed in the performance contract. Other types of paid-from- savings projects do not include a guarantee of savings and, therefore, will not require this type of rigorous energy audit. The investment-grade energy audit uses energy modeling to ensure the savings estimates are accurate. The audit is also used to help reveal whether the building system upgrades will meet the minimum level of energy performance required for LEED certification.

2. Basis for the M&V Plan

Data collected provides the information needed to develop the M&V Plan.

3. Basis for Preparing the Project Development Plan

Data collected provides details on the energy-saving opportunities and outlines the potential savings. It is also used to calculate the years needed to pay back the initial investment.

LEED Certification Assessment and LEED Credit Opportunities

Green PC projects add two tasks to the traditional performance contracting assessments— conducting a LEED certification assessment and identifying LEED credit opportunities.

The LEED certification assessment focuses on determining whether the building has the potential to meet the nine LEED prerequisites. The project team can conduct the assessment prior to the Green PC project process or include it as part of the ESCO's investment-grade energy audit. The audit also provides the opportunity to assess other building performance requirements needed for LEED certification. The ESCO will include the list of the LEED credit opportunities in the audit report.

The investment-grade energy audit can be paid for as a separate contract with the ESCO or, if the ESCO is chosen, the audit costs can be rolled into the Green PC Agreement. Information on Green PC Audit Agreement Requirements is in **Appendix I.**

The Project Development Plan is derived from the information collected during the investment-grade energy audit. It includes the list of project measures that will generate savings. Plans for Green PC projects will also include measures that do not produce savings, but are needed to achieve LEED certification. The final Project Development Plan for Green PC projects needs to bundle all the measures to ensure the desired ROI and simple pay-back period are met.

STEP A	STEP C	STEP E
Green PC Preparation	Investment-Grade Energy Audit	Measurement & Verification Plan

STEP B		STEP F
ESCO Partner Selection		Green PC Agreement

STEP D
Project Development Plan

Figure D-1 provides examples of Green PC project measures identified in an investment-grade energy audit and the cost and pay-back period related to each one. Longer pay-back measures are bundled with the quicker measures to create a project with a shorter overall pay-back period. **Figure D-1** also demonstrates how the ESCO can impact a number of LEED prerequisites and credits. Collaboration between the ESCO and the project team is critical to ensure LEED opportunities have been identified and those selected are included in the final Project Development Plan.

Figure D-1. Sample Project Development Plan – LEED Certification Opportunities

	MEASURE	COST	PAYBACK	CREDIT	BENEFITS
1	Replace existing roof with "cool roof" system	$15,000	11 years	SSc7.2	Reduces heat to microclimate of surrounding area and habitat. Helps to lower cooling load on a building's HVAC system.
2	Replace existing exterior lighting fixtures with full cut-off fixtures	$27,000	7 years	SSc8	Reduces light pollution to adjacent property and sky, which benefits nocturnal habitats and night sky viewing. Improved lighting efficiency of lamps reduces energy costs.
3	Install water submetering system	$7,000	N/A	WEc1	Allows building manager to track water use and identify opportunities for water efficiency improvements.
4	Install cooling tower water chemical management system	$11,000	9 years	WEc4.1	Reduces water use in cooling towers.
5	Replace boiler	$15,000	12 years	EAc1	Improves energy efficiency.
6	Replace pumps	$10,000	13 years	EAc1	Improves energy efficiency.
7	Install VSD's	$80,000	7 years	EAc1	Improves energy efficiency.
8	Upgrade lighting system	$250,000	6 years	EAc1	Improves energy efficiency.
9	Conduct an ASHRAE Level II Energy Audit	$12,000	N/A	EAc2.1	Identifies energy efficiency improvement opportunities.
10	Install system level metering	$19,000	N/A	EAc3.2	Allows building manager to track energy use and identify opportunities for energy efficiency improvements (exception management).
11	Test outside air ventilation rates	$14,000	N/A	IEQp1	Ensures that proper Outside Air ventilation rates are being maintained for all occupied spaces.
12	Install outside air delivery monitoring system	$37,000	N/A	IEQc1.2	Maintains healthy indoor air during occupied periods and alarms for conditions found outside of set parameters.
	Total Project	$497,000	8.1 years		

"N/A" for payback period is defined as not relevant to the project development plan due to not being included in the guarantee of savings. Potential for savings for many of those designated as N/A exists, but not to the degree that it can be practically predicted and easily measured in the context of guaranteed energy performance.

The Green PC Agreement is developed after the Project Development Plan is finalized. The Energy Services Agreement is the contract document for executing a traditional performance contract. It is the basis for the Green PC Agreement, but must be modified to include details on the efforts to seek LEED certification and to outline specific responsibilities of the ESCO and those of the owner. Since the ESCO guarantees the savings generated by the building system upgrades, the Green PC Agreement will articulate the conditions for reimbursement to the owner if the savings are not realized.

Depending on the ESCO's abilities, it could offer services to the owner to accomplish LEED credits, including ongoing tracking and monitoring, recordkeeping, performance measurement, training, and assistance in preparing the final LEED certification documents. ESCOs can provide such services as part of the agreement.

A sample Energy Service Agreement is available on the ESC Web site (www.energyservicescoalition. org). Sample language to modify the agreement for a Green PC project can be found in **Appendix J**.

Monitoring the performance of the building systems implemented to generate cost savings is needed to ensure the savings are realized. The elements of the M&V Plan are identified in the investment-grade energy audit, and its details finalized during the contract negotiations. Owners and ESCOs must thoroughly review the final M&V Plan before signing the Green PC Agreement.

1. M&V Plan

The M&V Plan outlines the procedures to measure the pre- and post-project conditions and to calculate the energy-use baseline.

2. Adjusted Baseline

The adjusted baseline is calculated the year when savings are actually realized. It is used to determine if the guarantee has been met.

The calculation method needs to be agreed to by both the owner and the ESCO and outlined in the M&V Plan. Details on calculating the adjusted baseline are in **Appendix G**.

3. Savings Guarantee

The ESCO and owner need to agree on how the guarantee is defined and calculated. For performance contracts, the definition includes not only the specific measures to be installed to improve building performance, but the specified level at which the measure must perform and whether savings can accrue from year to year. The investment-grade energy audit establishes the framework for calculating the guarantee.

M&V Plan Outline

- Detail the baseline conditions and data collected.
- Document all assumptions and sources of data.
- Determine what will be verified.
- Decide who will conduct the M&V activities.
- Create a schedule for all M&V activities.
- Discuss risk and savings uncertainty.
- Outline the details of the engineering analysis performed.
- Outline the details of the baseline energy and water rates.
- Describe the performance period adjustment factors for energy, water, and O&M rates, if used.
- Determine how energy and cost savings will be calculated.
- Provide details of any O&M cost savings claimed.
- Outline O&M reporting responsibilities.
- Document the content and format of all M&V reports, including post-installation commissioning, and the periodic and annual reports.
- Detail how and why the baseline may be adjusted.
- Define preventive maintenance responsibilities.

APPENDIX A
Glossary of Key Terms

Building Information Modeling (BIM)

BIM is a process to generate and manage building data over the lifecycle of a building. It is three-dimensional, real-time modeling software that engineers, architects, and ESCOs use to create data-rich models of existing facilities to predict the building's energy performance and to help determine the savings that can be generated from building system upgrades.

Bundling

In paid-from-savings projects, building system improvements generate utility cost savings. These savings are leveraged to help fund the project. Paid-from-savings projects seeking LEED certification can "bundle" or aggregate the utility cost-saving measures with non-cost saving measures to optimize green opportunities and project economics. When longer pay-back measures are combined with the quicker measures, the project will have a shorter overall pay-back period and higher ROI.

Energy Service Company (ESCO)

An ESCO is a business that develops, installs, and arranges financing for projects designed to improve the energy efficiency and maintenance costs for facilities. ESCOs generally act as project developers and assume the technical and performance risk associated with the project. Services are "bundled" into the project's cost and are repaid through the cost savings generated. The ESCO can identify and evaluate energy and water saving opportunities, provide engineering services from design to equipment specifications, order and install equipment, manage construction, provide long-term energy management and maintenance services, guarantee performance and savings, and, if needed, arrange financing.

Green Performance Contracting (Green PC)

Green performance contracting (Green PC) is based on the same project-delivery method as traditional performance contracting, but enhances the process by utilizing the LEED for Existing Buildings: O&M rating system as the criteria for a comprehensive green project.

Leadership in Energy and Environmental Design® (LEED)

The U.S. Green Building Council's LEED rating system is an internationally recognized certification system that measures how well a building or community performs across all the metrics that matter most: energy savings, water efficiency, CO_2 emissions reduction, improved indoor environmental quality, and stewardship of resources and sensitivity to their impacts. LEED provides building owners and operators a concise framework for identifying and implementing practical and measurable green building design, construction, operations, and maintenance solutions. The rating systems are designed for new and existing commercial, institutional, and residential buildings. Each rating system is organized into credit categories: Sustainable Sites, Water Efficiency, Energy and Atmosphere, Materials and Resources, Indoor Environmental Quality, and Innovation in Design. LEED points are awarded on a 100-point scale, and credits are weighted to reflect their potential environmental impacts. A project must satisfy all prerequisites and earn a minimum number of points to be certified. Certification levels, based on the number of points, include: Certified, Silver, Gold, and Platinum.

LEED for Existing Buildings: Operations & Maintenance Rating System

The LEED for Existing Buildings: O&M rating system is a set of voluntary performance standards for the sustainable, ongoing operation of buildings that are not undergoing major renovations. It addresses high-performance building systems, O&M best practices, and sustainable policies. The rating system can be applied both to existing buildings seeking LEED certification for the first time and to projects previously certified under LEED for New Construction, Schools, or Core & Shell. It is the only LEED rating system under which buildings are eligible for recertification.

Measurement & Verification (M&V)

Paid-from-savings projects with third-party financing need an added measure of assurance that system upgrades will perform. The M&V process includes specific methods and calculations to ensure the expected savings are realized. Costs are associated with the M&V process; therefore, it is essential to develop the appropriate measurement method for each cost-saving improvement. Performance problems can arise if savings are "under-measured" and M&V costs can be unnecessarily high if savings are "over measured", so striking a balance is needed. Projects that use performance contracting have the benefit of a guarantee from the ESCO that the cost savings from the system upgrades will be generated. An M&V Plan is implemented to ensure the guaranteed savings are realized.

Paid-from-Savings Funding Approach

The paid-from-savings approach is a financing strategy to green existing buildings. It leverages the savings generated from utility system upgrades to pay for a comprehensive greening project within a defined pay-back period. Paid-from-savings financing options include: self-financing, tax-exempt lease purchase agreements, power purchase agreements, performance contracts, equipment finance agreements, and commercial loans or bond financing for qualifying entities. In many cases, successful projects may employ a combination of these financing options, along with supplemental funding, such as revolving loan funds, utility rebates, renewable energy grants, and tax incentives. The paid-from-savings approach allows owners to implement needed repairs and upgrades, achieve reductions in energy and water use, and incorporate other green strategies and technologies in the most cost-effective manner.

Performance Contracting (PC)

Performance contracting, also known as energy performance contracting (EPC) or energy savings performance contracting (ESPC), is a well-established means of procuring and financing needed building repairs and upgrades. It is both a paid-from-savings financing strategy and a project-delivery method. PC projects implement project measures that focus on the building's physical systems that produce utility cost savings, such as energy and water system upgrades. The costs savings are guaranteed by the ESCO contracted to do the work. As with all paid-from-savings projects, the costs savings are leveraged to pay for the project within a defined pay-back period.

Project Measures

Project measures are upgrades, improvements, adjustments, and new or modified practices, policies, and programs included in the building's renovation plan. The types of project measures for the LEED for Existing Buildings: O&M rating system are: high-performance building systems, O&M best practices, and sustainability policies. The measures include implementing building improvements and technologies in order to use less energy, less water, and fewer natural resources; adopting O&M best practices to ensure project measures are effectively implemented and maintained; and establishing green policies, programs, and plans to demonstrate an organization-wide commitment to sustainability.

Return on Investment (ROI)

$$\frac{\text{annual net savings}}{\text{total cost}} = \text{ROI (\%)}$$

To calculate the ROI, divide the *annual net savings* by the *project's total cost*. The ROI will vary depending on both the size of the project and the amount of savings produced.

Simple Pay-Back Period

$$\frac{\text{total cost}}{\text{annual net savings}} = \frac{1}{\text{ROI \%}} = \text{simple pay-back (years)}$$

The simple pay-back period is the number of years it will take to recover installation costs based on annual cost savings. To calculate simple pay-back, divide the *project's total cost* by the *annual net savings*. The simple pay-back is the mathematical inverse of the ROI. To determine a target for simple pay-back, match the ROI to the life expectancy of installed building systems.

APPENDIX B
LEED Prerequisites and Credits and Their Savings Potential

The table below lists all prerequisites and credits in the LEED for Existing Buildings: O&M rating system. The prerequisites and credits are identified as either having a high potential to generate savings or having some potential to affect savings. Prerequisites and credits that yield no direct monetary savings are also noted. From the tables, owners can get a sense of the types of prerequisites and credits that help generate the savings required for a paid-from-savings project seeking LEED certification.

In most cases, the opportunities for savings are found in the prerequisites and credits related to building system improvements and upgrades. The final project measures for a paid-from-savings project seeking LEED certification will not be chosen solely for their potential to generate savings, but by other factors as well, including the overall project economics, financing, skill levels of the facilities and procurement staffs, state laws and regulations, the project provider's expertise, and the owner's overall sustainability goals.

SUSTAINABLE SITES	High Potential to Generate Savings	Potential to Impact Savings	No Direct Savings
SSc1 – LEED Certified Design and Construction		X	
SSc2 – Building Exterior and Hardscape Management Plan			X
SSc3 – Integrated Pest Management, Erosion Control, and Landscape Management Plan		X	
SSc4 – Alternative Commuting Transportation			X
SSc5 – Site Development – Protect or Restore Open Habitat			X
SSc6 – Stormwater Quantity Control			X
SSc7.1 – Heat Island Reduction – Nonroof		X	
SSc7.2 – Heat Island Reduction – Roof		X	
SSc8 – Light Pollution Reduction		X	
WATER EFFICIENCY	**High Potential to Generate Savings**	**Potential to Impact Savings**	**No Direct Savings**
WEp1 – Minimum Indoor Plumbing Fixture and Fitting Efficiency	X		
WEc1 – Water Performance Measurement	X		
WEc2 – Additional Indoor Plumbing Fixture and Fitting Efficiency	X		
WEc3 – Water Efficient Landscaping	X		
WEc4 – Cooling Tower Water Management	X		

ENERGY AND ATMOSPHERE	High Potential to Generate Savings	Potential to Impact Savings	No Direct Savings
EAp1 – Energy Efficiency Best Management Practices – Planning, Documentation, and Opportunity Assessment	X		
EAp2 – Minimum Energy Efficiency Performance	X		
EAp3 – Fundamental Refrigerant Management		X	
EAc1 – Optimize Energy Efficiency Performance	X		
EAc2.1 – Existing Building Commissioning – Investigation and Analysis		X	
EAc2.2 – Existing Building Commissioning – Implementation	X		
EAc2.3 – Existing Building Commissioning – Ongoing Commissioning	X		
EAc3.1 – Performance Measurement – Building Automation System	X		
EAc3.2 – Performance Measurement – System Level Metering	X		
EAc4 – On-site and Off-site Renewable Energy		X	
EAc5 – Enhanced Refrigerant Management		X	
EAc6 – Emissions Reduction Reporting			X

MATERIALS AND RESOURCES	High Potential to Generate Savings	Potential to Impact Savings	No Direct Savings
MRp1 – Sustainable Purchasing Policy			X
MRp2 – Solid Waste Management Policy			X
MRc1 – Sustainable Purchasing – Ongoing Consumables		X	
MRc2 – Sustainable Purchasing – Durable Goods		X	
MRc3 – Sustainable Purchasing – Facility Alterations and Additions		X	
MRc4 – Sustainable Purchasing – Reduced Mercury in Lamps		X	
MRc5 – Sustainable Purchasing – Food			X
MRc6 – Solid Waste Management – Waste Stream Audit			X
MRc7 – Solid Waste Management – Ongoing Consumables		X	
MRc8 – Solid Waste Management – Durable Goods			X
MRc9 – Solid Waste Management – Facility Alterations and Additions		X	

INDOOR ENVIRONMENTAL QUALITY	High Potential to Generate Savings	Potential to Impact Savings	No Direct Savings
IEQp1 – Minimum Indoor Air Quality Performance		X	
IEQp2 – Environmental Tobacco Smoke (ETS) Control			X
IEQp3 – Green Cleaning Policy			X
IEQc1.1 – IAQ Best Management Practices – IAQ Management Program			X
IEQc1.2 – IAQ Best Management Practices – Outdoor Air Delivery Monitoring		X	
IEQc1.3 – IAQ Best Management Practices – Increased Ventilation		X	
IEQc1.4 – IAQ Best Management Practices – Reduce Particulates in Air Distribution		X	
IEQc1.5 – IAQ Best Management Practices – IAQ Management for Facility Alterations and Additions		X	
IEQc2.1 – Occupant Comfort – Occupant Survey			X
IEQc2.2 – Controllability of Systems – Lighting		X	
IEQc2.3 – Occupant Comfort – Thermal Comfort Monitoring		X	
IEQc2.4 – Daylight and Views		X	
IEQc3.1 – Green Cleaning – High-Performance Cleaning Program		X	
IEQc3.2 – Green Cleaning – Custodial Effectiveness Assessment		X	
IEQc3.3 – Green Cleaning – Purchase of Sustainable Cleaning Products and Materials		X	
IEQc3.4 – Green Cleaning – Sustainable Cleaning Equipment		X	
IEQc3.5 – Green Cleaning – Indoor Chemical and Pollutant Source Control			X
IEQc3.6 – Green Cleaning – Indoor Integrated Pest Management		X	

APPENDIX C
Education and Information Resources

Organizations and agencies that offer relevant information, sample documents, and other resources for paid-from-savings projects are listed below.

Association of Energy Engineers (AEE)
www.aeecenter.org

The Association of Energy Engineers is a non-profit professional society of 9,500 members in 73 countries. AEE's mission is to promote the scientific and educational interests of those engaged in the energy industry and to foster action for sustainable development. AEE offers a full array of informational outreach programs, including seminars, conferences, journals, books, and certification programs. In 2002, the Efficiency Valuation Organization (EVO), formerly IPMVP Inc., in conjunction with AEE, established the Certified Measurement & Verification Professional (CMVP) program to raise the professional standards and improve the practice of those engaged in measurement and verification. AEE also conducts various workshops and Webinars on performance contracting and measurement and verification of performance contracting projects.

Building Owners and Managers Association (BOMA) International
www.boma.org/Resources/BEPC/Pages/default.aspx

The BOMA Energy Performance Contracting (BEPC) model is a model contract with supporting documents that allow building owners and operators to execute energy efficiency retrofits to existing buildings. BOMA and the Clinton Climate Initiative (CCI), in collaboration with major real estate companies and ESCOs, has developed a standardized, user-friendly contracting model that allows building owners and operators to successfully execute larger, more sophisticated retrofits and bring greater operational improvements in investment real estate.

Clinton Climate Initiative (CCI)
www.clintonfoundation.org/what-we-do/clinton-climate-initiative/

The Clinton Climate Initiative (CCI) Energy Efficiency Building Retrofit Program (EEBRP) brings together many of the world's largest cities, energy service firms and financial institutions in a landmark effort to reduce energy consumption in existing buildings. CCI works with industry, financial, government, and building partners to overcome market barriers and develop financially sound solutions that accelerate the growth of the global building efficiency market. CCI provides support to building owners such as city governments, commercial portfolio owners, schools, universities, and public housing authorities in identifying, designing, and implementing large-scale energy efficiency retrofit projects and brings the owner together with the necessary contracting and financial firms for implementation.

Database of State Incentives for Renewables & Efficiency (DSIRE)
www.dsireusa.org

The Database of State Incentives for Renewables & Efficiency Web site contains information on federal and state incentives, such as rebate and renewable energy grants.

Efficiency Valuation Organization (EVO)
www.evo-world.org

The Efficiency Valuation Organization began as the International Performance Measurement and Verification Protocol Committee, a group of volunteers who came together under a U.S. Department of Energy initiative to develop an international measurement and verification protocol to determine energy savings from energy efficiency projects in a consistent and reliable manner. In 2004, the non-profit corporation changed its name to the Efficiency Valuation Organization to reflect its expanded mission. EVO's Web site contains documents, industry news, resources, and discussion forums for online collaboration with colleagues around the world. A free copy of *IPMVP Volume I*, available for download, describes best practices for measuring and verifying savings from building improvement projects.

Energy Services Coalition (ESC)
www.energyservicescoalition.org

The Energy Services Coalition is a national non-profit organization composed of experts from a wide range of organizations working together at the state and local level to increase energy efficiency and building upgrades through energy savings performance contracting. Energy savings performance contracting enables building owners to use future energy savings to pay for up-front costs of energy-saving projects, eliminating the need to use their capital budgets.

On the ESC Web site, under "Resources and Info" (http://www.energyservicescoalition.org/resources/), are links to information on guaranteed energy savings performance contracting, state programs and activities, federal programs, financing, legislation, and model documents. Many states have ESC chapters that conduct informational meetings and workshops on how to implement performance contracting projects. Chapter contacts can be found on the ESC Web site.

EPA's ENERGY STAR® for Buildings and Plants
www.energystar.gov

The U.S. Environmental Protection Agency's ENERGY STAR program has many resources and tools to help building owners and facility managers improve the performance of their facilities. On the ENERGY STAR home page, find "Buildings and Plants," and open the link for "Tools and Resources Library." Additional information on Portfolio Manager, Cash Flow Opportunity (CFO) Calculator, and Target Finder can be found on the Web site.

Federal Energy Management Program (FEMP)
www1.eere.energy.gov/femp

The Federal Energy Management Program is part of the Department of Energy. Its Web site provides energy management resources and tools for federal agencies, some of which have a broader application at the state and local levels. Information is available by opening "Financing Mechanisms" and then "Super Energy Savings Performance Contracts."

National Association of Energy Service Companies (NAESCO)
www.naesco.org

The National Association of Energy Service Companies is a national trade association that promotes the benefits of energy efficiency. NAESCO works to help open new markets for energy services by directly promoting the value of demand reduction to customers through seminars, workshops, training programs, publication of case studies and guidebooks, and the compilation and dissemination of aggregate industry data. The NAESCO Web site has information on how to locate an ESCO, an overview of the latest technologies being used in performance contracting projects, case studies, and more.

National Clearinghouse for Educational Facilities (NCEF)
www.ncef.org

Created in 1997 by the U.S. Department of Education and managed by the National Institute of Building Sciences, the National Clearinghouse for Educational Facilities provides information on planning, designing, funding, building, improving, and maintaining safe, healthy, high-performance schools. Among the topics listed on the Web site are references on performance contracting, green schools performance features, power purchase agreements, and alternative financing.

U.S. Department of Energy's Guide to Financing EnergySmart Schools
www1.eere.energy.gov/buildings/energysmartschools/financing_guide.html

The new *Guide to Financing EnergySmart Schools* provides guidance on the process of financing energy efficient school renovations, retrofits, or new construction and outlines the advantages and disadvantages of a variety of financing mechanisms. Written for school administrators and school board members, the guide describes: principles of financing high-performance schools, such as determining project objectives, performing lifecycle cost analysis, and selecting a cost-benefit analysis method; making a business case for high-performance schools; non-energy benefits of energy-efficient schools; internal financing; debt financing; leasing arrangements; energy savings performance contracts; and financing information resources.

U.S. Green Building Council (USGBC)
www.usgbc.org

The U.S. Green Building Council is a 501(c) (3) non-profit community of leaders working to make green buildings available to everyone within a generation. The Leadership in Energy and Environmental Design (LEED) rating system is a third-party certification program and the nationally accepted benchmark for the design, construction, and operation of high-performance green buildings. LEED provides building owners and operators with the tools they need to have an immediate and measurable impact on their buildings' performance. Detailed information about greening existing buildings is available on USGBC's Web site.

Detailed information on greening existing buildings is available on USGBC's Web site. Copies of the LEED for Existing Buildings: Operations & Maintenance checklist and rating system are available for free download and the *LEED 2009 Reference Guide for Green Building Operations and Maintenance* is available for purchase. USGBC also offers a variety of LEED instructor-led workshops, online courses, and Webinars. To learn more about USGBC's LEED curriculum, visit www.usgbc.org/education.

APPENDIX D
LEED Certification Assessment

The LEED Certification Assessment form is used to determine if the building can meet the nine LEED prerequisites needed for LEED for Existing Buildings: O&M certification.

LEED CERTIFICATION ASSESSMENT SUMMARY

All prerequisites are met or can be met: ☐ Yes   ☐ No

PREREQUISITE	Is met:		Can be met:	
1. WEp1 – Minimum Indoor Plumbing Fixture and Fitting Efficiency	☐ Yes	☐ No	☐ Yes	☐ No
2. EAp1 – Energy Efficiency Best Management Practices – Planning, Documentation, and Opportunity Assessment	☐ Yes	☐ No	☐ Yes	☐ No
3. EAp2 – Minimum Energy Efficiency Performance	☐ Yes	☐ No	☐ Yes	☐ No
4. EAp3 – Fundamental Refrigerant Management	☐ Yes	☐ No	☐ Yes	☐ No
5. MRp1 – Sustainable Purchasing Policy	☐ Yes	☐ No	☐ Yes	☐ No
6. MRp2 – Solid Waste Management Policy	☐ Yes	☐ No	☐ Yes	☐ No
7. IEQp1 – Minimum Indoor Air Quality Performance	☐ Yes	☐ No	☐ Yes	☐ No
8. IEQp2 – Environmental Tobacco Smoke (ETS) Control	☐ Yes	☐ No	☐ Yes	☐ No
9. IEQp3 – Green Cleaning Policy	☐ Yes	☐ No	☐ Yes	☐ No

Notes:

Responsibility for Completion: ☐ ESCO   ☐ Owner   ☐ Other   This prerequisite can be met: ☐ Yes   ☐ No

WEp1 – MINIMUM INDOOR PLUMBING EFFICIENCY

Responsible Person:

Performance Period:

ASSESSMENT

1. a. Building has previously achieved a LEED-NC or LEED for Schools certification.   ☐ Yes ☐ No

   b. OR building was initially constructed in 1993 or later. (Construction documents are available for verification.)   ☐ Yes ☐ No

   c. OR, building plumbing fixtures were installed prior to January 1, 1993. (Additional data required)   ☐ Yes ☐ No

   d. OR building plumbing fixtures were installed prior to 1993. (See instructions for Performance Calculations Worksheet.)   ☐ Yes ☐ No

2. Describe the plumbing fixture and fitting inspection, testing, or preventive maintenance program in place for the project building. Describe how the inspection, testing, or program ensures the following:
   a. Flush and/or flow valves do not leak and
   b. Auto-flush and/or flow sensors are calibrated so fixtures flush and/or flow properly and at the proper time but not too often or for too long a time (i.e., generally only one flush per fixture use).

3. Have on hand a copy of the policy mandating an economic assessment of conversion to high-performance plumbing fixtures and fittings as part of any future indoor plumbing renovation (*required upload*).

Water Performance Calculation (See instructions for PERFORMANCE CALCULATIONS WORKSHEET)

4. Describe the inputs in the Fixture Group Definitions Table (the methodology used to define each fixture group, the derivation of data in each row, and gender ratios if the default is not used for any fixture group).

5. Enter flush fixture data for each fixture group defined in the Fixture Group Definitions Table. To account for dual-flush fixtures, enter a weighted average for GPF.

6. Have on hand manufacturer or supplier data to verify flow rates for each flush fixture and flow fixture type that differs from UPC/IPC efficiency requirements or obtain measured data on flush or flow rates for at least 20% (by number of fixtures) of each type.

Notes:

CREATING A WATER FIXTURE PERFORMANCE CALCULATIONS WORKSHEET

The *Fixture Group Definitions Table* below allows project teams to group all plumbing fixtures in the project together or separate them into sub-groups. Buildings that have the same usage patterns across all fixtures and have fixtures of similar efficiencies or vintages can be addressed in a single group. Otherwise, a different fixture group may be created for each unique usage pattern. For example, if fixture usage patterns are different on the first floor of the project building compared to other floors, create a fixture group for the first floor and a fixture group for all other floors. In some cases, it may be necessary to have a single user represented in more than one group in the table. No fixture may be represented in more than one group. Fixtures considered to be "Completely Replaced" involve replacements of all components that affect the gallons per use (i.e., for urinals and toilets it includes both the flush valve and the porcelain).

Fixture Group Definitions Table

Fixture Group	# of Fixtures Installed or Completely Replaced before 1993	# of Fixtures Installed or Completely Replaced in or after 1993	Annual Days of Operation	# of Full-Time Employees	# of Students

(Construct a table with these headings. Add a row for each fixture group.)

Flush Fixture Data Table

Fixture Group	Fixture ID (Optional)	Fixture Family[1]	Fixture Type[2]
Fixture Group A			
Fixture Group B			

(Construct a table with these headings and enter data for each fixture group.)

[1] Select from the following:
Water Closet
Urinal

[2] Select from the following:
For Water Closets: IPC/UPC Equivalent Water Closet, Low-Flow Water Closet, Ultra Low Flow Water Closet, Composting Toilet, Other
For Urinals: IPC/UPC Equivalent Urinal, Non-water Urinal, Other

Flow Fixture Data Table

Fixture Group	Fixture ID (Optional)	Fixture Family[1]	Fixture Type[2]
Fixture Group A			
Fixture Group B			

(Construct a table with these headings and enter data for each fixture group.)

[1] Select from the following:
Private Lavatory Faucet
Public Lavatory Faucet
Kitchen Sink
Shower

[2] **For Private Lavatories:** IPC/UPC Equivalent Lavatory, Low-Flow Lavatory, Ultra Low Flow Lavatory, Other
For Public Lavatories: IPC/UPC Equivalent Lavatory, Low-Flow Lavatory, Ultra Low Flow Lavatory, Other
For Kitchen Sinks: IPC/UPC Equivalent Kitchen Sink, Low-Flow Kitchen Sink, Other
For Showers: IPC/UPC Equivalent Shower, Low-Flow Shower, Other

The data in the three tables above will be needed to calculate the water efficiency performance of the project. Project teams can use the *LEED 2009 Green Operations & Maintenance Reference Guide* or the LEED Online submittal template (WEp1) to calculate the water efficiency percentage.

Notes:

Responsibility for Completion: ☐ ESCO ☐ Owner ☐ Other **This prerequisite can be met:** ☐ Yes ☐ No

EAp1 – ENERGY EFFICIENCY BEST MANAGEMENT PRACTICES: PLANNING, DOCUMENTATION AND OPPORTUNITY ASSESSMENT

Responsible Person:

Performance Period:

ASSESSMENT

Required Documentation

1. Building Operating Plan	☐ Yes	☐ No
2. Systems Narrative (see Systems Documentation Worksheet)	☐ Yes	☐ No
3. Sequence of Operations (see Systems Documentation Worksheet)	☐ Yes	☐ No
4. Narrative describing the building's preventive maintenance plan and schedule for the equipment described in the Systems Narrative.	☐ Yes	☐ No

ASHRAE Level 1 Walk-Through Analysis

5. Annual energy use breakdown by major end uses or applications (data table or graphical summary).	☐ Yes	☐ No
6. Energy Utilization Index using Portfolio Manager and comparison for potential cost savings. (Must be documented in submittal template for EAc1.)	☐ Yes	☐ No
7. List of potential low/no-cost changes, expected savings (kW and kWh), and maintenance cost savings.	☐ Yes	☐ No
8. Can all "No" items be resolved during the project?	☐ Yes	☐ No

Notes:

SYSTEMS DOCUMENTATION WORKSHEET

Systems documentation required for EAp1 includes a *Systems Narrative* and an excerpt of the *Sequence of Operations*.

Document:	Requirements:
Systems Narrative	**The following systems are addressed (at a minimum):** ☐ Space heating (e.g. group all boilers) ☐ Space cooling (e.g. group all chillers) ☐ Ventilation ☐ Lighting ☐ Building control systems **The following additional systems are addressed:** ☐ Domestic water heating ☐ Humidification and/or dehumidification **The narrative addresses the following aspects of the control system:** ☐ Central automatic ☐ Local automatic ☐ Occupant control **The narrative addresses the differences in system types:** ☐ Different floors ☐ Interior zones ☐ Perimeter zones **The narrative includes summaries of the following:** ☐ Central plant ☐ Distribution ☐ Terminal units (as applicable)

Sequence of Operations
A 1 to 2-page representative excerpt from the current Sequence of Operations for at least two different systems summarized in the Systems Narrative. The systems must be of different types (e.g., a chiller and a boiler).
	The excerpt addresses the following minimum content requirements, as applicable:
	☐ Detailed system-level documentation for each base building system
	☐ Desired operational states for various conditions in the building, including:
	   ☐ Which systems are running vs. idle
	   ☐ Whether operation is full-load or part-load
	   ☐ Staging or cycling of compressors, fans, or pumps
	   ☐ Proper valve positions
	   ☐ Desired system water temperatures
	   ☐ Duct static air pressures
	   ☐ Reset schedules or occupancy schedules in place
	   ☐ Information on operating phases, setpoints and controls, and feedback systems to monitor performance

Notes:

Responsibility for Completion: ☐ ESCO ☐ Owner ☐ Other **This prerequisite can be met:** ☐ Yes ☐ No

EAp2 – MINIMUM ENERGY EFFICIENCY PERFORMANCE

Responsible Person:

Performance Period:

ASSESSMENT

1. An ENERGY STAR Portfolio Manager Account has been properly established for the project building, including all energy fuels: | ☐ Yes ☐ No

2. Performance Determination Methods:

a. Case 1: Projects Eligible for ENERGY STAR Rating | ☐ Initial ENERGY STAR rating: _____
☐ A rating has not been obtained.

b. Case 2, Option 1: Projects Not Eligible for ENERGY STAR Rating | ☐ Initial adjusted benchmark score: _____
☐ A score has not been obtained.

c. Case 2, Option 2: Projects Not Eligible for ENERGY STAR Rating | ☐ Initial alternative score: _____
☐ A score has not been obtained.

3. The minimum energy efficiency performance has been met. | ☐ Yes ☐ No

4. Potential rating/score after project completion: | Projected rating/score: _____

5. Project has the potential to meet the minimum energy efficiency performance requirement. | ☐ Yes ☐ No

6. Describe how the projected rating was determined:

Notes:

Responsibility for Completion: ☐ ESCO ☐ Owner ☐ Other This prerequisite can be met: ☐ Yes ☐ No

EAp3 – FUNDAMENTAL REFRIGERANT MANAGEMENT

Responsible Person:

Performance Period:

ASSESSMENT

1. Building uses CFC-based refrigerants:	☐ Yes ☐ No
2. If yes, refrigerant conversion or system replacement can be done with a simple payback of 10 or fewer years:	☐ Yes ☐ No

Notes:

| Responsibility for Completion: ☐ ESCO ☐ Owner ☐ Other | This prerequisite can be met: ☐ Yes ☐ No |

MRp1 – SUSTAINABLE PURCHASING POLICY

Responsible Person:

Performance Period:

ASSESSMENT

1. Project has an environmentally preferable purchasing policy.	☐ Yes ☐ No
2. If yes, policy includes all the USGBC required elements.	☐ Yes ☐ No
3. Policy addresses ongoing consumables (at a minimum: paper, toner cartridges, binders, batteries, and desk accessories).	☐ Yes ☐ No
4. Policy addresses at least one of the following additional purchasing categories (check all that apply): ☐ Durable goods (furniture, electrical appliances), or, ☐ Facility alterations and Additions, or, ☐ Mercury containing lamps	☐ Yes ☐ No

Notes:

Responsibility for Completion: ☐ ESCO ☐ Owner ☐ Other **This prerequisite can be met:** ☐ Yes ☐ No

MRp2 – SOLID WASTE MANAGEMENT POLICY

Responsible Person:

Performance Period:

ASSESSMENT

1. Organization has a solid waste management (recycling) policy/contract.	☐ Yes ☐ No
2. If yes, policy/contract includes all the required elements.	☐ Yes ☐ No
3. Policy addresses reducing, reusing, or recycling of all waste categories noted below (check all that apply): ☐ Ongoing consumables ☐ Durable goods ☐ Facility alterations and additions ☐ Mercury-containing light bulbs	☐ Yes ☐ No

Notes:

IEQp1 – MINIMUM INDOOR AIR QUALITY PERFORMANCE

Responsible Person:

Performance Period:

ASSESSMENT

Air Handling Units:	AHU-1	AHU-2	AHU-3	AHU-4	AHU-n
1. Constant Volume (CV) or Variable Air Volume (VAV):	☐ CV ☐ VAV	☐ CV ☐ VAV	☐ CV ☐ VAV	☐ CV ☐ VAV	☐ CV ☐ VAV
2. AHU provides outside air flow required by 62.1-2007:	☐ Yes ☐ No	☐ Yes ☐ No	☐ Yes ☐ No	☐ Yes ☐ No	☐ Yes ☐ No
3. Minimum required outside air flow (CFM):					
4. Measured outside air flow (CFM):					
5. Date of measurement:					
6. AHU Compliance?	☐ Yes ☐ No	☐ Yes ☐ No	☐ Yes ☐ No	☐ Yes ☐ No	☐ Yes ☐ No
7. If ASHRAE 62.1 ventilation rates cannot be met, can a rate of at least 10 CFM/person be met?	☐ Yes ☐ No				

8. Describe the measurement method (measurement device, accuracy, and how measurements are taken).

9. For VAVs describe how the VAV outside air flow is set up during the air flow measurements to operate at the worst-case condition expected during normal operations (i.e., fan speeds set at minimum normal operating level, OA dampers set at their minimum normal operating opening, etc.).

10. Describe the ventilation maintenance program, including a description of the periodic checks and scheduled maintenance performed. Describe whether the checks are manual, based on a building automation system, or both.

11. If a building automation system is used for any ventilation components, have on hand a periodic system maintenance status report taken during the performance period which covers those components. If any ventilation components are handled manually, have on hand a copy of the maintenance log written during the performance period which covers those components (*required upload*).

12. Have on hand documentation verifying implementation of a preventive maintenance program during the performance period (*required upload*).

13. Project team will perform or oversee tests on all local, dedicated exhaust systems within the project building during the performance period to confirm proper function. Have on hand at least one testing report for each separate type of exhaust system (*required upload*).

Notes:

Responsibility for Completion: ☐ ESCO ☐ Owner ☐ Other **This prerequisite can be met:** ☐ Yes ☐ No

IEQp2 – ENVIRONMENTAL TOBACCO SMOKE (ETS) CONTROL

Responsible Person:

Performance Period:

ASSESSMENT

1. a. State or local statutes or regulations govern smoking.	☐ Yes ☐ No
b. OR, the building has a smoking policy that bans smoking in the building/grounds.	☐ Yes ☐ No
c. OR, a previous LEED certification governs ETS Control.	☐ Yes ☐ No

Notes:

Responsibility for Completion: ☐ ESCO ☐ Owner ☐ Other **This prerequisite can be met:** ☐ Yes ☐ No

IEQp3 – GREEN CLEANING POLICY

Responsible Person:

Performance Period:

ASSESSMENT

1. Organization has a Green Cleaning Policy.	☐ Yes	☐ No
2. Custodial operation is outsourced to a private contractor.	☐ Yes	☐ No
3. Policy/contract includes all the required elements.	☐ Yes	☐ No
4. Policy addresses all of the following topics noted below: (Select all that currently apply.)	☐ Yes	☐ No

☐ Guidelines for purchase of sustainable cleaning and hard floor and carpet care products meeting the sustainability criteria outlined in IEQc3.3. (The project team is not required to apply for IEQc3.3, but·the Green Cleaning Policy must adhere to the requirements of the credit.)

☐ Guidelines for purchase of cleaning equipment meeting the sustainability criteria outlined in IEQc3.4. (The project team is not required to apply for IEQc3.4, but the Green Cleaning Policy must adhere to the requirements of the credit.)

☐ Standard operating procedures (SOPs) addressing how an effective cleaning and hard floor and carpet maintenance system will be consistently utilized, managed, and audited. If applicable, it specifically addresses cleaning to protect vulnerable building occupants.

☐ Strategies for promoting and improving hand hygiene, including both hand washing and the use of alcohol-based waterless hand sanitizers.

☐ Guidelines addressing the safe handling and storage of cleaning chemicals used in the building, including a plan for managing hazardous spills or mishandling incidents.

☐ Requirements for staffing and training of maintenance personnel appropriate to the needs of the building. Requirements address the training of maintenance personnel in the hazards of use, disposal, and recycling of cleaning chemicals, dispensing equipment, and packaging.

☐ Provision for collecting occupant feedback and continuous improvement to evaluate new technologies, procedures, and processes.

Notes:

APPENDIX E
LEED for Existing Buildings: O&M Policy Model

Policy Elements Guidance

USGBC provides the following policy model to help the project team ensure that all relevant and necessary aspects of the policy are addressed. (Note: this policy model appears in the LEED for Existing Buildings: O&M rating system; for reference it is repeated here without change.)

Policy Model

Any policies required by the LEED for Existing Buildings: O&M rating system must, at a minimum, contain the following components:

Scope – Describe the facility management and operations processes to which the policy applies. Describe the building components, systems, and materials to which the policy applies.

Performance Metric – Describe how performance will be measured and/or evaluated.

Goals – Identify the sustainability goals for the building.
Note: although applicants are required to set goals, documentation of actual achievement is not required to demonstrate compliant policies; stating the goal is enough. Applicants are encouraged to set high goals and work toward their achievement.

Procedures and Strategies – Outline the procedures and strategies in place to meet the goals and intent of the policy.

Responsible Party – Identify the teams and individuals involved in activities pertaining to the policy. Identify and outline key tasks for the above teams and individuals.

Time Period – Identify the time period over which the policy is applicable.

Applicants are not required to develop separate policies for the purposes of achieving LEED for Existing Buildings: O&M prerequisites and credits as long as they can highlight these components in their existing policies.

APPENDIX F
Financing Options

Third-Party Financing

If third-party financing is needed, the organization should start the financing planning early in the process. Some ESCOs may be able to suggest financing alternatives, or you may prefer to retain the services of a financial advisory firm specializing in energy, water and LEED projects, especially if the project is large and/or complex.

There are many financing vehicles and structures that can be used to underwrite the certification process. In general, tax-exempt financing offers lower interest rates than taxable financing, so determining if your organization is qualified to issue tax-exempt obligations should be the first step. Any U.S. state or district which has the power to levy taxes, police power, or the right of eminent domain, qualifies. Such entities include state agency buildings, public school facilities, state universities and community colleges, public hospitals, libraries, fire departments, and town halls. If the organization does not qualify directly, it may be eligible by using a conduit agency, such as a dormitory or health authority or community development corporation. The conduit agency will, however, charge a fee.

Tax-Exempt Lease-Purchase Agreements

A tax-exempt lease-purchase agreement is an effective alternative to traditional debt financing because it allows organizations to pay for energy-saving upgrades by using money set aside in the annual utility budget. When properly structured, tax-exempt lease-purchase agreements make it possible for public sector or qualifying organizations to draw on the anticipated savings from future utility bills to pay for new, energy-efficient equipment and related services up front.

In most states, a tax-exempt lease-purchase agreement does not constitute a long-term "debt" obligation because of non-appropriation and/or abatement language written into the agreement, which may mean that public approval is not required. Non-appropriation language effectively limits the payment obligation to the organization's current operating budget period, typically 12 months. The organization will, however, have to assure lenders that the energy-efficiency upgrades being financed are considered of *essential use*, which minimizes the non-appropriation risk to the lender. If, for some reason, future funds are not appropriated, the equipment is returned to the lender, and the repayment obligation is terminated at the end of the current operating period without placing any obligation on future budgets. Abatement language limits the payment obligation to the ability to use the equipment and may be required in some states.

Qualifying organizations should consider using a tax-exempt lease-purchase agreement to pay for energy-efficiency equipment and related services when the projected savings will be greater than the cost of the equipment (financing costs included). While the financing terms for tax-exempt lease-purchase agreements may extend as long as 20 to 25 years, they are usually less than 15 years and are limited to the useful life of the equipment.

Commercial Loans, Equipment Leases, and Equipment Finance Agreements

If the organization cannot issue tax-exempt obligations, commercial loans, equipment leases, and equipment finance agreements are other options. These methods impact the organization's balance sheet differently, and are subject to various approval processes and credit requirements.

Commercial Loans

Options include: amortized and balloon or "bullet" loans. An amortized loan payment includes both principal and interest components and seeks to pay down the principal to zero over the life of the loan. A balloon or "bullet" loan payment consists mostly of interest costs, with little or no principal pay down. Balloon or "bullet" loans have a large payment due at the end of the loan term, which covers the repayment of the outstanding principal balance.

Equipment Leases

Equipment leasing is often referred to as "creative financing." Under an equipment lease, the lender owns the equipment and leases it to the organization for a defined period of time for a set cost. At the end of the lease term, the organization may purchase the equipment at its fair market value or for a predetermined price. It can also continue the lease, lease new equipment, or return the equipment. From an accounting perspective, leases fall into two categories: operating or capital leases. In order to qualify for operating lease (off balance sheet) treatment, the lease must meet FASB 13 requirements, which means that the equipment being financed must retain a strong residual value. Lease payments can be structured to match your cash flow needs, resulting in step up or down payments, skip payments, etc.

Equipment Finance Agreements (EFA)

An equipment finance agreement, also known as a conditional sales agreement, is an agreement for the purchase of an asset in which the borrower is treated as the owner of the asset for federal income tax purposes, thereby being entitled to the tax benefits of ownership, such as depreciation, but does not become the legal owner of the asset until the terms and conditions of the agreement have been satisfied.

Power Purchase Agreements (PPA)

A power purchase agreement is a contract between an electricity generator and a building owner to provide electricity at guaranteed rates. This is used for renewable energy projects like solar and wind. The electricity generator is usually a utility company and is referred to as the PPA provider.

The owner purchases energy from the PPA provider. The PPA provider in turn secures funding for the project, maintains and monitors the energy production, and sells the electricity to the owner at a contracted price for the term of the contract. The term usually runs ten to 25 years. In some contracts, the owner has the option to purchase the generating equipment from the PPA provider at the end of the term. Other options may include renewing the contract with different terms or having the equipment removed. Insurance on the system may be provided by the PPA provider.

A PPA allows the building owner to utilize renewable energy without making a large up-front capital expenditure. The owner is able to lock-in an energy rate over the term of the contract, resulting in significant cost savings. In addition, a PPA gives a tax-exempt entity, such as a school, non-profit, or government agency, the opportunity to take advantage of federal tax incentives for renewable energy. By assigning system ownership as well as all rebates and tax credits to the PPA provider, the owner is able to reduce the system's installation costs significantly, resulting in a lower rate for the owner.

Supplemental Funding

Supplemental funding is a category of financing that does not incur a formal recurring obligation or debt repayment. Examples are utility rebates, renewable energy grants, and revolving funds.

Utility Rebates and Renewable Energy Grants

Utility rebates or renewable energy grants are used to reduce capital costs, which in turn reduce the amount of financing needed, making the project more attractive to lenders.

Revolving Funds

Many institutions, including state governments and universities, have established revolving funds to finance building improvements that generate utility cost savings. The savings, in turn, are used to replenish the fund.

APPENDIX G
Measurement & Verification

The Efficiency Valuation Organization (EVO) and the U.S. Department of Energy have developed guides and resources to address M&V best practices. The International Performance Measurement and Verification Protocol (IPMVP), maintained by EVO, is the industry standard.

Savings Determination

Calculating the *value* of actual savings is a two-step process. First, the actual savings must be determined. This can be measured by utility units, such as kilowatt hours (kWh) and demand (kW). Second, the units must be assigned a value.

Step One – Determine Actual Savings

Graphs G-1 and **G-2** track a company's annual energy use, showing energy-conservation measures installed in Year-4 and energy use declining in Year-5.

> **Question:** What is the actual decline in energy use from Year-4 to Year-5?
> **Answer:** One may think the actual reduction in energy use is determined by comparing Year-5 usage to Year-4 usage. See **Figure G-1**.

However, the more accurate calculation requires determining an adjusted baseline for Year-5 and using that number to assess the actual reduction in energy use. See **Figure G-2**.

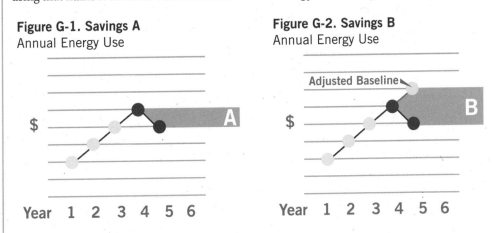

Figure G-1. Savings A
Annual Energy Use

Figure G-2. Savings B
Annual Energy Use

Since the adjusted baseline becomes the basis of the savings, it is important to know how it is calculated and what factors are considered. A graph of annual energy use for any organization is not as simple as usage depicted in **Figures G-1** and **G-2**. The graphs show a steady increase in annual energy use, possibly due to added loads and/or building square foot increases. In reality, the increases — or decreases — vary from year to year, so the calculation of the adjusted baseline will be more complicated.

Step Two – Determine the Value of the Savings

The next step is to determine the value of the savings, also known as savings valuation. Marginal cost is a charge for a given unit of measure. Electrical energy is complex because there are separate rates for consumption (kWh) and demand (kW). The savings that occur for consumption and demand should be calculated separately and then added together. A common mistake for valuing electrical energy savings is to use the average unit cost method, also known as a blended rate, which is determined by taking the total cost of the bill and dividing it by consumption (kWh) to determine the new level of consumption.

> **The following statement is from the IPMVP, Volume 1, paragraph 8.1.2:**
>
> "Average, or blended prices, determined by dividing billed cost by billed consumption, is often different from *marginal costs*. In this situation, average prices create inaccurate statements of cost *savings* and should not be used."

Using the Average Unit Cost Method

The following example demonstrates the problem caused by using the average unit cost method instead of calculating demand and consumption separately.

An owner implemented a lighting control program to reduce energy use. Average unit cost, which was always used for developing electrical energy budgets, was used to value the units of energy (kWh) saved. The program, which consisted of encouraging occupants to turn off the lights when leaving, saved a significant amount of energy. However, the savings occurred during off-peak hours. The owner wanted to value the energy reduction using the average unit cost method.

The cost equates to multiplying $0.126/kWh ($13,691/108,333 kWh) by the amount of kWh saved. The problem with this calculation method is the other marginal rate component of the bill (Demand, kW) was not affected by the energy reduction effort since it occurred during off-peak hours. Therefore, the actual rate for kWh, $0.09 ($0.07 energy charge + $0.02 fuel charge) should have been used to multiply the amount of kWh saved. Using average unit cost in this case inflated the reported savings by 40%.

ABC Electric Company

Bright Elementary School
Account # 345F3907AA
Meter # 068-3448
Billing Period 09-10-08 to 10-09-08

Service Charge	$150
Energy [kWh]	108,333
Energy Charge (.07/kWh)	$7,583
Fuel Charge (.02/kWh)	$2,167
Demand [kW]	516
Demand Charge ($4/kW)	$3,265
Tax	$526
Total	$13,691

APPENDIX H
Green PC RFP/Q Language

For most Green PC projects, the RFP/Q process is used to select an ESCO. Some states maintain a list of qualified ESCOs from which public sector organizations can choose. The RFP/Q for a Green PC project needs to convey to the marketplace the owner's desire to achieve LEED for Existing: Operations & Maintenance certification.

Sample RFP/Q language for a Green PC project is outlined below.

- **[Organization's name]** seeks to achieve USGBC's LEED for Existing Buildings: O&M certification.
- The successful ESCO will be acquainted with USGBC's LEED for Existing Buildings: O&M rating system and certification process.
- At least one member of the ESCO project team must be a LEED Accredited Professional (LEED AP).
- ESCOs should provide details of their involvement with all LEED for Existing Buildings or LEED for Existing Buildings: O&M certified projects to date.
- A representative from the ESCO will be required to participate on the owner's project team, which will include six, two-hour meetings during the course of the Green PC project.
- The ESCO will perform a LEED certification assessment for the project prior to or in conjunction with the investment-grade energy audit. See Attachment **[X]** for a copy of the LEED Certification Assessment form. The tasks to be conducted by the ESCO are marked accordingly.
- The ESCO will be required to use industry-standard energy modeling software to develop an energy-use baseline. The ESCO will also be required to use EPA's Target Finder to determine the feasibility of improving the building's energy performance rating to the level needed for LEED for Existing Buildings: O&M certification (ENERGY STAR rating of at least 69). The results of this feasibility study will be included in the audit report.
- The investment-grade energy audit will be conducted such that the requirements of the ASHRAE Level I and Level II Energy Audits are met.
- Construction and equipment installation activities will conform to the requirements of the LEED for Existing Buildings: O&M rating system as follows:
  - Materials and Resources Credit 3–Sustainable Purchasing — Facility Alterations and Additions

- Materials and Resources Credit 9–Solid Waste Management — Facility Alterations and Additions
- Indoor Environmental Quality Credit 1.5–IAQ Best Management Practices — Management for Facility Alterations and Additions

Owners wishing to include BIM as a technical strategy for developing the Green PC project should also include the following language:

- The successful ESCO will know how to use BIM and integrate its use throughout the Green PC process.
- The ESCO will conduct a BIM-based LEED certification assessment prior to or in conjunction with the BIM-based investment-grade energy audit. See Attachment **[Y]** for a copy of the LEED Certification Assessment form.
- The ESCO will be required to run a building energy model using industry-standard BIM software and gbXML Schema to develop an energy-use baseline. As a part of the process, the ESCO will be required to use an energy analysis service to determine the feasibility of improving the building's energy efficiency performance to the level needed for LEED for Existing Buildings: O&M certification. The results of this feasibility study will be included in the investment-grade energy audit report.
- Ownership of all data inputs and information outputs associated with the BIM model will be retained by the Owner.

APPENDIX I
Green PC Audit Agreement

Under traditional performance contracting, the ESCO will complete an investment-grade energy audit to include an analysis of each proposed measure and the projected energy and cost savings related to it. The ESCO will also propose terms for the agreement and present a proposal that includes recommended measures, financing terms, and projected annual cash-flow analysis.

For Green PC projects, the owner may want to add the following requirements to the investment-grade energy audit to assist with LEED certification:

Investment-Grade Energy Audit Report

- The ESCO will include in the investment-grade energy audit report, as a separate enclosure, the results of the audit that meet the requirements of an ASHRAE Level I Energy Audit. (See Energy and Atmosphere Prerequisite 1 – Energy Efficiency Best Management Practices — Planning, Documentation, and Opportunity Assessment.)

- The ESCO will include in the investment-grade energy audit report, as a separate enclosure, the results of the audit that meet the requirements of an ASHRAE Level II Energy Audit. (See Energy and Atmosphere Credit 2.1 – Existing Building Commissioning — Investigation and Analysis.)

- The ESCO will perform the tasks detailed in the LEED Certification Assessment Form that have been designated as the responsibility of the ESCO.

- The ESCO will use EPA's Target Finder to determine the building's site energy use intensity needed to achieve the minimum ENERGY STAR rating of 69. For projects not "ratable" under the ENERGY STAR system, the ESCO will determine the related energy performance level by using USGBC's "Case 2 Calculator", which can be found under the "Registered Project Tools" section of USGBC's Web site.

- The ESCO will include in the investment-grade energy audit report a description of how the list of proposed building system upgrades will attain the required energy performance rating (ENERGY STAR rating of at least 69). If no combination of upgrades will yield the required energy performance rating, the audit report will note as such.

Project Development Plan

- Based on guidance from the owner, the ESCO will include in the Project Development Plan, a list of measures identified as "Savings Opportunities". These are building improvement measures that will generate cost savings and may help achieve LEED credits as designated by the owner.

  Note: Even though performance requirements for LEED credits may be met by these measures, there may be additional requirements that must also be met in order to earn the credit. Elements like plans, schedules, or verification documentation may also be required to satisfy the credit.

- Based on guidance from the owner, the ESCO will include in the Project Development Plan, a list of measures identified as "LEED Certification Opportunities". The list will include project measures that will help accomplish the LEED credit requirements designated by the owner.

APPENDIX J
Green PC Agreement Language

The Green PC Agreement is a complex document that brings together all aspects of the performance contracting process. A sample Energy Services Agreement, which is used as the basis for a Green PC Agreement, can be found on the Energy Services Coalition's Web site.

A major section of the agreement will contain a list of "Schedules" that correspond with key aspects of the agreement. The list below is from the sample Energy Service Agreement found on the ESC Web site. It is typical of the types of schedules found in traditional performance contracts. A Green PC Agreement, however, will need to modify several of the schedules. Explanations for the modifications are outlined below. (Note: The letters and the titles of schedules are not universal; they will vary from contract to contract. Those listed are representative and provide a sense of the types required for a Green PC project.)

Schedule A – Equipment to be Installed by ESCO
Include any specific equipment features or installation methods required to accomplish LEED certification.

Schedule B – Description of Premises; Pre-Existing Equipment Inventory

Schedule C – Energy and Water Cost Saving Guarantee

Schedule D – Compensation to ESCO for Annual Services
Address any LEED certification services to be provided by the ESCO.

Schedule E – Baseline Energy Consumption

Schedule F – Savings Measurement and Verification Plan; Methodology to Adjust Baseline

Schedule G – Construction and Installation Schedule
Address any requirements for recycling or demolishing old equipment. Special installation procedures, such as IAQ *Guidelines for Occupied Buildings Under Construction*, from the Sheet Metal and Air Conditioning National Contractors Association (SMACNA) should be noted.

Schedule H – Systems Start-Up and Commissioning; Operating Parameters of Installed Equipment
Include commissioning needs identified in the LEED for Existing Buildings: O&M rating system, such as the Energy and Atmosphere Credit 2.3 – Existing Building Commissioning — Ongoing Commissioning.

Schedule I – Standards of Comfort

Schedule J – ESCO's Maintenance Responsibilities

Schedule K – Institution's Maintenance Responsibilities

Schedule L – Facility Maintenance Checklist

Schedule M – ESCO's Training Responsibilities

Could contain training and assistance for other LEED credits accomplished by the owner.

Schedule N – Installment Lease and Payment Schedule

Schedule O – Alternative Dispute Resolution Procedures

Schedule P – Final Project Cost and Project Cash Flow Analysis

Schedule Q – LEED Certification Project Credits

This schedule is unique to the Green PC Agreement and is not included in a traditional Energy Services Agreement, the traditional performance contracting agreement. Schedule Q lists the Green PC project's LEED prerequisite and credit requirements and includes any special instructions to the ESCO to ensure the requirements are met. The schedule should clearly define the LEED prerequisites and credit requirements, if any, that are the responsibility of the ESCO and the prerequisites and credit requirements under the purview of the owner.

APPENDIX K
BIM Overview

Building information modeling is a process to generate and manage building data over the lifecycle of a building. It is three-dimensional, real-time modeling software that engineers, architects, and ESCOs use to create data-rich models of existing facilities to predict the building's energy performance and to help determine the savings that can be generated from the building system upgrades. BIMs can be created in the pre-project planning phase by owners or in the pre-proposal stage by architects, engineers, or ESCOs.

In the pre-proposal stage, project teams should work with architects, engineers, or ESCOs to create the BIM, which captures key information about the facility, including location, building type, square footage, room volumes, occupancy capacity, equipment locations, weather data, and building envelope. The exercise will highlight possibilities and provide a model to be used throughout the building's lifecycle.

To determine the new ENERGY STAR rating resulting from the proposed building system upgrades, the BIM data will need to be exported to an energy analysis tool. Below is an example of how the process is implemented and is representative of the steps required.

BIM data is exported to The Green Building XML (www.gbxml.org), which utilizes Web-based energy analysis tool (ex: Green Building Studio) to report, not only the new ENERGY STAR rating resulting from the proposed building system upgrades, but information such as:

- Water Usage and Costs Evaluation
- Energy and Carbon Results
- Whole Building Energy Analysis
- Photovoltaic Potential
- Daylighting
- Weather Pinpointing and Detailed Weather Analysis
- Carbon Emission Reporting
- Wind Energy Potential
- Natural Ventilation Potential

The same model used to benchmark can be used for the investment-grade energy audit and again during the project's implementation stage. In addition, the BIM can also assist in verifying savings if Option D of the IPMVP is used.

Note: If either an Energy Scoping BIM or BIM-based investment-grade energy audit was completed, an owner must be clear in both the RFP/Q and audit agreements as to the ownership of the BIM. The BIM asset is valuable to the project under consideration, and if proceedings are delayed, the owner can reuse the model at a later time.

BIM Resources

gbXML – The Green Building XML Schema
http://www.gbxml.org/about.htm

Highly accurate BIMs of existing facilities can be exported to this industry-standard green building format allowing it to share important building information with eQUEST, DOE-2 and EnergyPlus. The Green Building Schema (gbXML) is central to utilizing BIM throughout a facility's lifecycle.

U.S. General Services Administration's (GSA) BIM for Energy Performance
http://www.gsa.gov/bim

Executive Order 13123 is a national initiative to reduce the average annual energy consumption of the GSA's building inventory. With the use of BIM, GSA hopes to strengthen the reliability, consistency, and usability of predicted energy-use and energy-cost results. Specific benefits to a project team may include: more complete and accurate energy estimates earlier in the process, improved lifecycle costing analysis, increased opportunities for measurement and verification during building occupation, and improved processes for gathering lessons learned in high-performance buildings.